Effective Approaches to Human Ecology Education

EDUCATIONAL CONTRIBUTIONS IN ETHNOBIOLOGY

EDUCATIONAL CONTRIBUTIONS IN ETHNOBIOLOGY

Marsha Quinlan, Justin M. Nolan, and Sarah Walshaw, Series Editors

Educational Contributions in Ethnobiology is a peer-reviewed monograph series dedicated to original book-length publications in ethnobiology and related social and natural sciences. The volumes, rather than present research, present methods, theory, and pedagogy that advances ethnobiological research.

Becoming a Researcher
Steve Wolverton

Effective Approaches to Human Ecology Education
Daniela J. Shebitz, Elizabeth A. Olson, and, Steve Wolverton, Editors

Effective Approaches to Human Ecology Education

Edited by

Daniela J. Shebitz, Elizabeth A. Olson, and Steve Wolverton

Society of Ethnobiology

2021

Dedication

Justin Nolan came into our lives as an anthropologist but will forever be our *Oganali* (friend). His kind and generous spirit set the stage for a unique and close relationship with the Cherokee people with whom he worked and played in the wild woods of the Cherokee Nation's Flint and Goingsnake Districts.

Justin cared deeply for the Cherokee people and worked with local Elders, traditional artists, and Cherokee language first speakers to record and interpret their traditional ecological knowledge into the twenty-first century. Through his years of collaborative field work with rural Cherokees in Oklahoma, Justin revolutionized local Native peoples' concept of "ethno" as it applied to their own understanding of the biological sciences.

Justin's gift to the Cherokee people was his dedication to teaching locals to recognize and identify their own indigenous knowledge systems as still intact and still practiced in today's modern Cherokee society, as well as demonstrate how to better record, analyze, and perpetuate this knowledge for future generations. Those Cherokee people who worked closely with Justin will be forever grateful for knowing him.

Do-na-da-go-hv-i, until we meet again.
Roger & Shawna Cain

Shawna Cain, Cherokee National Treasure
Roger Cain, Cherokee National Treasure

Library of Congress Control Number: 2021920999

ISBN 978-0-9990759-6-8 (paperback)
ISBN 978-0-9990759-7-5 (PDF)

Society of Ethnobiology
Boston University Archaeology Room 345,
675 Commonwealth Ave., Boston, MA 02215

Cover photo: Roger Cain, Cherokee National Treasure, shows Justin Nolan a patch of bloodroot on his ancestral lands in rural Adair County, Oklahoma. Nolan is an ethnobiologist at the University of Arkansas, where Roger is pursuing graduate studies in anthropology. Photo courtesy of Roger and Shawna Cain.

Table of Contents

List of Figures

List of Tables

Effective Approaches to Human Ecology Education[1]

Daniela J. Shebitz[2], Elizabeth A. Olson[3], and Steve Wolverton[4]

Introduction

The interdisciplinary nature of human ecology as a field of study lends itself to unique collaborative learning opportunities that can occur both in- and out-of the classroom. The silo-structured education that is typical in higher education is not effective, for the students nor the faculty. Yet as college educators, we have found that it is all too easy to fall into habits of incorporating methods of instruction that do not challenge our students to become creative and to recognize the importance of integrating skills and subjects. As ethnobiologists, many of us have been trained through an integrative approach to learning, holding degrees in a variety of disciplines including anthropology, geography, biology, linguistics, and other disciplines that can fall within the vast field of human ecology. Correspondingly, our classes on human–environment interactions are taught from a variety of perspectives and vary in content. And, while we may be the lone voice of ethnobiology at our institutions, we thrive off of the conversations that are generated at our annual meetings with like-minded peers. This edited volume in the *Educational Contributions in Ethnobiology* series was in fact one such outcome of a Society of Ethnobiology conference. We recognized first that there was no great reference with ideas that incorporates interaction and collaborative learning in courses on human ecology. Second, we saw that we had some really terrific experience and expertise to pull from!

The following year, we organized a special section at the Society of Ethnobiology Conference that brought together peers with various backgrounds to share examples of class exercises that illustrate the interactions between humans and the environment so that these tools may be available for use by other ethnobiologists. The work presented not only covered various fields, but also ranged from first-year to senior level courses. It was the assemblage

1 This volume is a product of an organized session, by the same title, at the Society of Ethnobiology Annual Meeting, which took place June 3–7, 2018, in Madison, Wisconsin. The session was organized and co-chaired by Daniela Shebitz and Steve Wolverton.

2 Kean University

3 Southern Utah University

4 University of North Texas, Denton

that we were hoping it would be, and that we now share with you so that you may use it as an inspiration for your own courses.

The contributions are framed around several central tasks and questions: 1) Define the purpose of the course and how the course is situated in the broader university goals, 2) Explain the learning outcomes that were intended during design of the project or exercise, 3) Describe the exercise, including its context, content, structure, and requirements, 4) Share what aspects of the design worked well, which aspects were problematic, and how were those problems solved? 5) Finally, reflect on the outcomes of the learning activity from both student and pedagogical perspectives.

This volume presents a compilation of novel exercises that transcend the typical classroom format so that the education of students in ethnobiology can be enriched through unique learning experiences. The contributions are arranged by intended course level, beginning with first-year and general education courses, and proceeding up through advanced courses and field projects. Owing to the interdisciplinary nature of ethnobiology, the activities described in this volume are well-suited for a variety of college and university courses.

The volume opens with a non-traditional model of a General Education curriculum that is offered to first-year students at Southern Utah University. Through this Jumpstart program, students complete all of the required 28 GE credits through an interdisciplinary year-long course that focuses on human ecology and "Our National Parks." The unique learning activities presented by Dr. Liz Olson and Dr. William Heyborne are transferrable not only for institutions that are looking to develop integrative courses, but also for stand-alone classes that focus on human ecology or ethnobiology.

Another unique exercise for college freshmen is shared by Dr. Denise Glover in Chapter 2. She presents an activity that focuses on the American crow where students in the first-year seminar course develop skills of observating and recording data that connects to information they read from Marzluff and Angell's *In the Company of Crows* (2005). Their deliverable for that assignment is an article for the campus newspaper that draws people's attention to crows as other than pests and incorporates their own observations as well as those from scholars. Since crows are commonly found on campuses across the country, we hope that you find this assignment easy to incorporate into your own curriculum.

As human ecologists and ethnobiologists, we recognize that, as Drs. Fidji Gendron, Vincent Ziffle, and Nametha Srikanthan present in Chapter 3, "participation of Indigenous Elders in science courses can transform education and provide learning opportunities for all students." Yet many of us may wonder how to effectively incorporate local knowledge and stories into a college science classroom. The authors share how participation of Indigenous Elders has enhanced Biology and Chemistry courses, and the benefits that the students gained by learning from these Elders. In particular, hands-on laboratory activities (hide tanning, making hand lotion with native plants, and creating porcupine quill bookmarks) were developed under the guidance of an Elder and were remarkably effective at helping students to understand the natural world better.

Dr. Amanda Thiel presents a one-day experiential, outdoor, medical ethnobotany workshop, that she has presented to groups of local university students over the past decade. The workshop is an introduction to local plants and their medicinal uses in the Western U.S. herbal tradition, and instructs in the methods and rationales of preparing an herbal extract from plants harvested throughout the educational plant walk. In Chapter 4, Dr. Thiel discusses what does and does not work for adapting this exercise for outdoor education.

Once students have moved on to their upper level courses, they bring with them not only knowledge from their foundational classes, but also from an increasing variety of "real life" experiences that they have gained during their undergraduate education. A number of the chapters in this volume focus on the students' upper level coursework and invite them to contribute their growing knowledge to classroom activities.

Two chapters focus on educational initiatives at the University of North Texas. First, Dr. Steve Wolverton explains a semester-long activity in a seminar course that he adapted from David Haskell's book, *The Forest Unseen: A Year's Watch in Nature*, for ethnobiology. Mandala exercises were established for students so that they would take part in routine visits to small outdoor places, to record observations of biota, place, and people (including themselves) and to note changes over time. Wolverton presents weekly explorations and assignments that allow students to describe experiences within their mandala that transform their observation and communication skills while connecting them to a single location. Next in Chapter 6, Dr. Lisa Nagaoka presents "Using the Management of Urban Species to Teach Conservation Biogeography" that was developed for an upper-level course in Applied Biogeography. Students take part in a mapping project that documents the distribution of native trees on campus and evaluates the possible reasons for their current distribution. This exercise challenges them to think as a historical ecologist based on their observations of their own campus and lessons from class. The class project consists of a blog and a brochure, both of which present formats that are easily adaptable to other courses.

Another activity for upper-level students is presented in Chapter 7 by Dr. John Richard Stepp from the Department of Anthropology at the University of Florida. In his ethnobiology course, he developed an exercise that leads students to realize that habitats that have been disturbed by anthropogenic activity often have higher numbers of medicinal or edible plants than intact forest habitats. In this activity, students work in small groups and are assigned either anthropogenic ("disturbed") habitats or forested habitats. They collect or identify every plant species within a pre-defined unit and then consult literature and field guides to determine if the plants within the area are medicinal and/or weedy. The results foster discussion about the strong link between useful plants and environmental conditions in human-modified landscapes.

As online learning is increasing in popularity throughout the world, Drs. Amanda Thiel and Dr. Marsha Quinlan present an example of how to offer an online lesson in Chapter 8, *Culture, Ecology, and Health in Mesoamerica.* In addition to sharing material that focuses on traditional medicine and ecological knowledge in Mesoamerica, they reflect on pedagogical theories relevant to online learning and teaching. This overarching perspective draws critical attention to factors that influence student success and satisfaction with online lessons.

We close this volume with an example of an experiential learning senior capstone course with the objectives of encouraging students to develop their ability to address and analyze complex problems and to creatively apply their acquired knowledge and skills to real-life situations and connecting students to the greater community. While Dr. Daniela Shebitz and Dr. Mike Edelstein present this model as adapted to public institutions in New Jersey, the model presented of empowering upper-level undergraduates to work with community partners is applicable to multiple fields and projects.

By creating this volume, it is our hope that we inspire our readers to creatively draw on their environment and to connect with their local communities when teaching and learning about human ecology. The following chapters are merely a sample of the great approaches to human ecology education that are already being practiced at colleges and universities. We hope that this sample will stimulate the development of many new techniques that will inevitably enrich the perspectives and knowledge for the next generation of ethnobiologists.

Chapter 1.

Jumpstart, Our National Parks: Using Local Resources and High Impact Educational Practices to Teach Human Ecology in Integrated General Education

Elizabeth A. Olson[1] and William H. Heyborne[1]

Chapter Summary

Southern Utah University (SUU), trademarked as the University of the Parks, is a flagship partner-institution with the National Park Service due to our proximity to, and history of engagement in, our region's parks and protected areas. Through a unique general education offering, known as Jumpstart, some first-year students at SUU are given the opportunity to complete all of their general education requirements in a single year-long integrated course. The course provides 28 credits, and is co-taught by up to eight faculty from a range of disciplines, including biology, geology, political science, writing, visual art, the humanities, and anthropology. Ethnobiology emerges at center-stage in this integrated general education course, due to its broad overlap with each of the other disciplines. The purpose of this chapter is to consider some of the high impact practices incorporated in teaching human ecology from an interdisciplinary perspective. Using examples of specific course activities, and reflecting on student attitudes and beliefs expressed during class discussions and in assignments, we suggest the learning activities described here can help faculty to have a meaningful impact on undergraduate students' perception of anthropogenic environmental impact. We give specific suggestions of integrated, lower-level learning activities that can facilitate understanding the complexity of human-environment interactions.

1 Southern Utah University

Introduction

Higher education has been moving toward interdisciplinary approaches to learning and teaching because of the advantages for student learners, along with the real-world applicability of interdisciplinary problem-solving. Fields such as human ecology and ethnobiology are easily taught in interdisciplinary settings given that the content deals with human-environment interactions. Hunn has pointed out that one of the key advantages of ethnobiology is learning to "love" the species that are studied which is, he argues, an essential step towards biodiversity conservation (2014). With the growth of scholarship in the field of ethnobiology (e.g., Hidayati et al. 2015; Hunn 2007) there is a complimentary need to train students in ways that will facilitate their inclusion in academic research and scholarship of integrated human-environment systems. Wolverton has dubbed this a move to "Ethnobiology 5"—the next phase of ethnobiological work (2013). We agree that there is an increasing need to train students in ethnobiological thinking, and that doing so will increase their love and appreciation for biodiversity. With this broader context in mind, we will consider an example of a special experiential general education course at Southern Utah University.

Southern Utah University

Our course is taught at Southern Utah University (SUU), a regional, public university serving approximately 12,000 students. SUU is located in southern Utah, near many national parks, protected areas, and publicly managed lands. Geographic proximity, combined with a long history of collaboration with land-management agencies, has led to the name *University of the Parks*. This moniker has been embraced by the University community. Consequently, the combination of our location and our outdoor-oriented student body create unequaled opportunities to teach human ecology and sustainability principles through place-based learning.

Southern Utah University, along with all public colleges and universities in Utah, has adopted the Association of American Colleges and Universities LEAP (Liberal Education and America's Promise) Challenge (Hawkesworth 2017). Endorsing carefully constructed Essential Learning Outcomes and research-verified High-Impact Educational Practices, such as first-year programs, learning communities, service learning, and community-based learning, both the faculty and administration at SUU are committed to the ideals of this initiative.

A Carnegie Community Engaged Institution, SUU is committed to helping our students understand the value of service to ones' community as well as the need for life-long learning. Southern Utah University is also a leader in experiential education, being home to the Experiential Learning Leadership Institute and the School of Integrative and Engaged Learning (SIEL). Part of this engaged learning mission has been the creation and orchestration of several sections of a unique team-taught general education course called Jumpstart. Jumpstart is a year-long, completely integrated, team-taught course, where students complete the entirety of their general education requirements. Each Jumpstart section is organized around a unifying theme. This paper represents experiences from the section entitled *Jumpstart: Our National Parks.*

Jumpstart

Southern Utah University offers three sections of Jumpstart each year. Capitalizing on our regional strengths in the performing arts, including the Utah Shakespeare Festival and a vibrant film industry, we offer a Jumpstart section entitled *Stage and Screen.* The second section, called *Sports, Health, and Identity,* appeals to students interested in a wide variety of majors and programs at SUU and works to accommodate the variable schedules of student athletes. The hallmark, and original, Jumpstart section which has been offered each year since 2015 is called *Our National Parks.* Organizationally, the goal of the Jumpstart general education program is to give students 28 college credits during one academic year. These 28 credits satisfy all of SUU's general education requirements. During that year, 40 students in each section work with a team of six to eight unique faculty from disciplines across campus to navigate through the completely integrated and team-taught lessons, with a focus on High-Impact Educational Practices (Kuh and Schneider 2008). The one non-integrated course is math. Due to students' variation in math readiness, Jumpstart students work with their advisor to select the most appropriate mathematics course to take during spring semester with the general student body. What this means is that the same 40 students are together for a shared learning experience from August through May. Likewise, they have the same faculty members during that entire academic year. This familiarity comes with strengths and challenges. We have found that the students really bond, both with each other and their faculty team, which leads to increased authenticity in the classroom. This can contribute to a greater degree of engagement in some cases. Drawbacks include unnecessary off-topic conversations during class time and some strains on the student-faculty divide, which is necessary for teaching undergraduate students.

Jumpstart: Our National Parks met Monday through Friday from nine o'clock in the morning until noon. Because it was a team-taught course, there were various faculty in the classroom at any given time—but only rarely were all of us in the classroom at the same time. Typically, an individual faculty member would schedule an hour at a time, one or more times a week, during which he or she would instruct in content specific to their discipline. Other faculty would join as available, or when specific instructional or content needs made this necessary. We incorporated multiple modes of instruction, including: lectures, seminars, discussions, laboratories, and multiple field trips. Because of the focus on the national parks, we took as many excursions into natural resource areas around our campus as possible. Emphasizing place-based experiential learning, the faculty team led educational trips to many local areas including: Cedar Breaks National Monument, Bryce Canyon National Park, Zion National Park, Grand Canyon National Park, Death Valley National Park, Grand Staircase–Escalante National Monument, Joshua Tree National Park, and the Hualapai Reservation. These field experiences occurred overnight and during the school week, typically, and were counted towards the "contact hours" since faculty and students were engaging in various learning and growth activities throughout each day. Due to SUU's proximity to these sites, and partnerships with many of them, students were given a once-in-a-lifetime opportunity to learn about human ecology and sustainability within a multidisciplinary, place-based context.

Initial Perceptions

An important element in understanding the impact and effectiveness of human ecology education is to consider the initial perceptions of both the faculty and the students regarding the course. This was especially important for us, as prior experience suggested that there may be a divide between faculty and students regarding not only the reasons students signed up for the course, but also the philosophical underpinnings of some of the course learning outcomes, especially those related to sustainability and human environmental impacts. What we discovered was that although faculty hoped the students were initially drawn to Jumpstart due to the integrated curriculum and innovative pedagogies, along with a passion for the national parks and outdoors, most students were actually drawn to Jumpstart simply because of the ease with which they could satisfy the general education requirements for whichever degree program they choose at SUU. Earning 28 or more credits during their first year, without the hassles associated with individual course scheduling, and the integrated teaching element, made it much easier for most of the students to check-off general education degree requirements. A secondary reason that students expressed for signing up for Jumpstart, was the National Parks theme for the course. Many, but not all, of the students came from backgrounds that included significant time spent in the outdoors, and were drawn to our section due to the possibility of furthering those avocations.

In terms of perceptions of human environmental impacts, most of the faculty came into this course with the expectation to, and even excitement to, teach sustainability and environmental problem-solving. Most of the faculty had prior experience relating our discipline to ecological issues. What we discovered regarding the students is that although they enjoyed nature and being outdoors, our students tended to associate "environmentalism" with a very specific, liberal, political orientation. Many of our students were brought-up in the conservative state of Utah, and during class discussions, students expressed that they viewed science as being sometimes biased and as a tool that is often used to further political agendas. What this meant for us in the classroom was that not all of the students were inclined to participate in conversations or learning about topics such as climate change, sustainability, or even recycling and environmental stewardship. This was challenging because the faculty came into the course with a passion for teaching about ecological and environmental issues—wanting to engage in problem-solving learning activities. As a faculty concerned about the future of our planet, there were challenges getting some students in the course to become engaged with the topics that they viewed as "political." Yet, there were some students in the class who identified with the environmental and sustainability issues, and we certainly felt the rift between those who did and those who did not. In classroom discussions, this was sometimes challenging. As faculty, we worked diligently to understand the different perspectives and beliefs held by students and tried to help students do the same.

Our goal in this chapter is to share some examples of what worked, and some of what did not work, regarding opening up dialog and overcoming barriers related to sustainability issues, so that we can contribute to more effective approaches to human ecology education within higher education. We will share some of the learning activities we used in Jumpstart as examples of the ways that High-Impact Educational Practices can be combined with place-based education to capitalize on area resources while achieving general education learning outcomes.

Description of the Learning Activities

Where Do You Stand?

One of the first activities we did at the start of the semester, to help establish a baseline for students pertaining to their own beliefs with regards to sustainability, and how their beliefs and opinions are formed, is the "Where do you stand?" activity. In this exercise, students are asked to "vote with their feet" regarding their agreement or disagreement with prepared statements about sustainability issues. We conducted this activity during one of our first field trips, so that the students would have an opportunity to reflect on nature and their place in nature. To start, we had the students stand out in a grassy field and they were told they had a spectrum of responses to the prepared statements. At one end of the field there was a tree and that spot represented complete disagreement with the statement; at the opposite end of the field was a point that signified complete agreement with the statement. The students were then asked to place themselves somewhere along the spectrum of agreement to disagreement with the various statements, such as: "With more than 7 billion people commuting from one place to another, drilling for oil is necessary to fuel more cars." After the students aligned themselves somewhere on the spectrum of complete agreement to complete disagreement, selected students were asked to share their perspective and justify their position. Interestingly, this conversation often caused some of the students to reevaluate their position, and thus move to a new place along the continuum. Some other sample questions that were used include: "Science and technology will ensure that food production and energy supplies keep up with the demands of the growing population"; "With the human population over 7 billion we must protect endangered species habitats by not developing them for any reason"; or "Automobile makers in this country should be required by law to make all of their vehicles fuel efficient even though the vehicles may become more expensive." This was a good activity for the students because it not only served as an icebreaker, but was also an opportunity for them examine their own perspectives and beliefs, to think about how those perspectives and beliefs are different from those of their classmates, and how their perspectives may have been shaped. Additionally, this provided the faculty with some information regarding what existing knowledge the students had, the sources of that knowledge, and how that knowledge had shaped their attitudes and beliefs about the world we live in and the role of humans in that system.

Timelines

Another learning activity that we implemented early in the semester was working with timelines. Given the diverse array of disciplines represented in our faculty team, we wanted to set the tone and define for the students the different disciplinary perspectives that we each come from. The disciplines that were represented during this learning activity included: geology, biology, anthropology, and political science. Our learning objective was to help learners develop an understanding of the different timescales of the various disciplines and to provide context for anchoring student experience related to time. This activity was conducted as an in-class exercise, and served as a springboard for a small group research project which included a poster and class presentation. The timeline activity itself was rather simple: we handed the students a set of cards. Each card had an event specific to one of the disciplines. We started with geology/biology. We put a piece of masking tape along the length of the room on the floor. The tape was then divided up into periods of time—again, specific to the discipline. The students were assigned to small groups and tasked with putting the discipline-specific events in order on the timeline. We began with the most expansive of the timelines (geology/biology) and then moved from discipline to discipline narrowing the timeline each time. For example, the timeline for geology/biology was 4.6 billion years—the age of Earth. Next came anthropology, with a timeline of 2 million years—the basis of human evolutionary history. We ended with political science, and the history of American Democracy—with a timeline of roughly 300 years.

The goal of the activity was to more firmly anchor the students into time by talking about the actual timing of the assigned event, as well as the difference in time between the various disciplines. This discussion led seamlessly, during the political science discussion, to the more recent history regarding protected areas, including National Parks—the theme of the entire course. Through this activity, students not only got a sense of the disciplinary scales, but also the impact of the scales on their own perspective. We felt this was important to establish early on in the semester because time and scale was a unifying theme throughout the entire course. For example, later in the course, we took students to the Grand Canyon (which took geological time to form) where we were able to relate that to the relatively small scale of political time, including the establishment of Grand Canyon National Park, or anthropological time, and the evolution of humans—or even the time humans have been living in or visiting that area.

Following this introductory in-class activity, the students were assigned to small groups and each group was given a major geologic time period to research. The assignment was to focus on the region of the planet that was to become the Grand Staircase–Escalante National Monument and illustrate via a poster presentation what was going on in this region in regards to life, land, water, and climate during their assigned time period. The posters were printed in full scale and hung on the walls of the classroom, where they served as a reference point for the rest of the year. We discovered that having a project of this magnitude, that necessitated working in small groups and doing research early on in the semester, set the tone regarding classroom expectations: A strong pedagogical tool.

Service Learning

We want to give an example of how service learning was incorporated into the course because service learning can aid in personal paradigm shifts. We wanted the service learning to focus on natural resources, human ecology, and local connections. We were fortunate to be able to collaborate with the Hualapai tribe and work on a habitat restoration project, on tribal land, inside the Grand Canyon. This was facilitated through a partnership and collaboration with an ethnobiologist who works in the Hualapai Tribe Department of Cultural Resources and partners with the Grand Canyon National Park to promote sustainability and protection of the natural and cultural resources in the area (Cannon 2017). The Hualapai Tribe invited our class group to participate in a cross-cultural exchange with another Indigenous group who were visiting them during the same time as our visit, and the students were able to learn dances, songs, and foods from the Hualapai—thereby enriching their knowledge of the significance of this type of work and the importance of collaboration with local cultural groups. Our task was to help remove invasive species manually (especially tamarisk, *Tamarix ramosissima*) and subsequently assist in planting native species (such as coyote willow, *Salix exigua,* which has both material and medicinal purposes for the Hualapai peoples) along the Colorado River (Cannon 2017). While engaged in the physical labor, students also had time during our two days there to reflect on the work they were doing and the significance that it holds for different groups of people as well as for the environment. As a result of their oral reflection in our group, and their conversations with Hualapai tribal members and others on the trip, some students reported experiencing a paradigm shift. Students' attitudes about conservation, native peoples, and human ecology began to change. This extra service learning trip came towards the end of the spring term and over a weekend, thus we did not require attendance but encouraged students to make time in order to participate. Students who volunteered for the service learning project noted how important they felt that experience was for their learning. They were able to have a cultural exchange with the Hualapai Tribe, learn from the ecologist and botanists working on the project, and actually see a physical transformation of the place occur. They saw that they had the power to not only learn about environmental problems, but to also actually work to help solve them. Putting learning into practice was a very empowering learning experience for the students.

ePortfolio Creation

At the end of the year, students were required to complete an ePortfolio which was comprised of "artifacts" that could be drawn from any of their various learning experiences throughout the year, throughout all of the courses, and from any of the various faculty who had been involved in Jumpstart. The challenge was for them to relate those different artifacts to the themes of sustainability and how humans interact with their environments. They organized their ePortfolios along three main axes: 1) social and cultural, 2) ecological and environmental, and 3) economic and political. In each of these sections they had to provide not simply a narrative of what the course learning activities were, but actually *reflect* on what they gained from that learning activity, or what they think they should have gained from it. This process of reflection is intended to reinforce learning. It is potentially very effective in a team-taught, integrated learning course such as this because it forces students to engage in critical thinking as they express what they have learned. ePortfolios have a broader impact because, "The LEAP Challenge calls on colleges and universities to build pathways where all undergraduates complete a substantial cross-disciplinary project in a topic *significant to the student and society,* as part of the expected pathway to a degree" (Hawkensworth 2017). While perhaps not of a level to be considered a substantial cross-disciplinary project, certainly, the ePortfolio is a high impact practice that helps students to practice this type of metacognitive learning.

Reflection

Using the strong foundation established through our course learning activities, we have recognized some important successes of the Jumpstart program overall. First, having the students actively engaged in small group projects meant they stayed engaged and did not become easily distracted by each other or outside influences during in-class work time. While not mentioned above, we frequently and routinely changed the composition of the groups. This did two things: it reduced formation of cliques or exclusion of some individuals, and it helped to break down barriers that may have existed between students due to differing backgrounds or perspectives. Second, working with local experts such as park rangers and the tribal experts added to the richness of the experience for the students. It also contributed to the rigor and the real-world application of the teaching and learning. Third, service learning in the field was among the richest learning experiences the students had because they felt the combination of service, academics, and the application of knowledge while connecting with people. Finally, student ePortfolios built throughout the year helped to reinforce learning and reminded students that while we may have been having fun, they still learned a lot!

As we reflect on the year-long Jumpstart course, however, we also pause to reflect on what could have been better. Number one, it might have been nice to expand the topics of the course to include themes beyond the national parks. Some students commented that because they were not very interested in outdoor recreation and national parks, they felt that the topic became tiresome and perhaps less interesting and engaging for them as the year progressed. Number two, students felt like they needed more concrete knowledge of what was expected of them upfront. While we as a faculty looked at the open nature of the course, in terms of projects and assignments, as an opportunity to evolve and be flexible as the course went on, students found that approach difficult to stay on top of, and sometimes lost track of their progress as learners. This may also be an asset of the course because it gives undergraduates the opportunity to transcend the structured learning environments common in many classrooms and become comparatively independent problem solvers—even though this process requires a period of transition that can be uncomfortable for them at the time. Finally, we learned that less is more. Looking back, we would have preferred to consolidate some of our interdisciplinary projects, so that as a faculty team we were less likely to overwhelm students with so many big projects. For example, faculty can team up on some of the projects and work together with the students on the different disciplinary facets of them. Although this was our intention, and we tried hard to adhere to that goal, it is also a fact that students are taking many threads of the course from many faculty simultaneously—despite the interdisciplinary nature of the course.

Adapting Jumpstart Learning Activities

We recognize that the year-long integrative general education course model may not be easily adaptable at many other institutions, but there are elements of the Jumpstart program that can be extrapolated and applied in other institutional settings. That is why our primary goal in this chapter has been to share concrete learning activities that we feel could be easily adapted into human ecology or ethnobiology courses being taught at other institutions. We believe that the "Where do you stand," service learning, and ePortfolio activities can be directly incorporated in other college courses without much challenge. The timelines activity, however, was directly linked to the interdisciplinary and team-taught structure of the Jumpstart course. This may make its incorporation into a more traditional college course a challenge. However, we have a couple of suggestions regarding how this might be incorporated into courses at other institutions.

One of the hallmarks of a twenty-first-century education is the acknowledgement of interdisciplinary collaborations and learning. We recognize that working with individuals from different disciplinary perspectives characterizes much of the real-world and this is a reason why we work to demonstrate this for our students. Team-teaching is an ideal way to accomplish this type of interdisciplinary learning. If team-teaching is not a possibility, due to scheduling or contractual constraints, the faculty member can consider inviting guest speakers from other disciplines to help facilitate the timelines activity. Finally, if unable to pursue either of these options, a faculty member may be able to use field trip experiences, and the guides or interpreters at important geological sites are usually equipped with sufficient knowledge of the specific place to be able to accomplish a modified version of the activity.

The long-range impacts of teaching ethnobiology from interdisciplinary perspectives are not yet certain, but we expect that there will be a positive impact on the ways that students will be able to comprehend the dynamics of environmental problems. Interdisciplinary problem-solving will be essential for the next generation who are already coping with impacts related to climate change. As "Ethnobiology 5" (Wolverton 2013) grows, students and professors who embark on interdisciplinary collaborations will be situated to lead in our understandings of the environmental problems we encounter.

References Cited

Cannon, C. 2017. TEK and Vegetation Restoration in Western Grand Canyon. Poster Presented at the 40th Annual Meeting of the *Society of Ethnobiology*, Montreal, QC, Canada, May 10–13, 2017. Available at: https://ethnobiology.org/conference/abstracts/40/tek-and-vegetation-restoration-western-grand-canyon. Accessed on January 5, 2021.

Hawkesworth, K. 2017. The LEAP Challenge. Available at: https://www.aacu.org/leap-challenge. Accessed December 20, 2018.

Hidayati, S., F.M. Franco, and R.W. Bussmann. 2015. Ready for Phase 5 - Current Status of Ethnobiology in Southeast Asia. *Journal of Ethnobiology and Ethnomedicine* 11:17. DOI:10.1186/s13002-015-0005-7.

Hunn, E. 2007. Ethnobiology in Four Phases. *Journal of Ethnobiology* 27(1):1–10.

Hunn, E. 2014. To Know Them is to Love Them. *Ethnobiology Letters* 5:146–150. DOI: 10.14237/ebl.5.2014.297.

Kuh, G.D., and C.G. Schneider. 2008. *High-impact Educational Practices: What They Are, Who Has Access to Them, and Why They Matter.* Association of American Colleges and Universities, Washington, DC.

Wolverton, S. 2013. Ethnobiology 5: Interdisciplinarity in an Era of Rapid Environmental Change. *Ethnobiology Letters* 4:21–25.

Chapter 2.

Caw Connections: Observing & Writing About Crows in College

Denise M. Glover[1]

Chapter Summary

As part of a first-year seminar course titled *People, Plants, and Animals*, I designed a module centered on learning about a common resident on many college campuses across the country: the American crow (*Corvus brachyrhynchos*). Students read selections from Marzluff and Angell's *In the Company of Crows and Ravens* (2005) and then perform their own observations of crows on campus, recording these observations. After group discussions about these two sources of data, students are assigned to write an article for the campus newspaper drawing people's attention to crows as other than pests. They must incorporate their own observations as well as those from scholars (Marzluff and Angell 2005, 2012), and are given the instruction to write compellingly. The project excites important functions in the development of college students, including skills of observation and recording, awareness of local surroundings and non-human beings living in those surroundings, and careful reading and writing that connects various types of data with their own intellectual and emotional experiences. It is one of the most highly valued projects in the class.

Introduction to the Institution and the Course

The University of Puget Sound is a four-year liberal arts college of approximately 2600 students located in a suburban area of the city of Tacoma, Washington. In 2018–2019, 27% of

1 University of Puget Sound

university students identified as students of color, 17% were first-generation college students, and 59% of students identified as female; additionally, a majority of students are from California or elsewhere on the West Coast. Annual tuition (not including room and board) at Puget Sound in 2018–2019 was $49,510. In short, many students come from affluent white families, and a majority of them were raised in urban or suburban settings. For these reasons, bringing awareness to key social and environmental issues of justice and stewardship are foundational in my pedagogical decision-making.

I have been teaching anthropology and Asian Studies at the University of Puget Sound since 2008.[2] In 2014, I began teaching a course that I designed for our Seminar in Scholarly Inquiry (previously known as Freshman Seminar) series, a sequence of two courses required of all first-year and transfer students that focuses on core issues of college scholarship. The course I designed is titled *People, Plants, and Animals* (with credit given to Eugene Hunn, as he taught a course with a similar name at the University of Washington when I was his student). Essentially what I did was to revamp my higher-division Ethnobiology course (titled *Environmental Anthropology* for my first few years of teaching at University of Puget Sound), reducing material and adding in specific assignments and readings that attended to various aspects of college scholarship, including but not limited to: identifying primary, secondary, and tertiary sources; identifying peer-reviewed scholarship and writing, particularly in contrast with popular writing; discussing writing structure, use of evidence in writing, and thesis construction (through identifying in required readings, but also in writing); identifying voice and use of literary devices in scholarly and popular writing; and how to tackle difficult readings with numerous reading strategies.

My goal with the class is to teach the topic of ethnobiology through a course focused on learning how to read carefully and critically, and to write effectively, using evidence and sometimes emotive force. Teaching about environment, stewardship, and human-non-human ecological interactions, and helping students develop a sense of awareness of their natural surroundings through a first-year seminar is an effective way to enhance sensibilities of these issues in students for two key reasons. First, the promotion of ecological literacy is integrated into the college experience and, hopefully, college identity and consciousness. Rather than segregating in discipline-specific (and usually upper-division) courses, ecological literacy and development is integrated with foundational college skills. Second, since this is not an elective course, students that might not have first thought of registering for such a course, or who plan on majoring in a non-related field, are exposed to the issues of environmental stewardship; thus, there is a wide target audience.

2 I wrote the final version of this chapter in summer 2019. Delays in publishing have resulted in a publication date of 2021. I have left the language throughout as I originally wrote it in 2019, although there have been several key personal and historical changes in this two-year span—including that while still affiliated with the University of Puget Sound, I am not currently teaching there.

Arguments of integrating science into core, non-science curricula have been made by the AAAS (2011) and Holm et al. (2010) in terms of biology, Harrison (2014) in terms of botany, and Wagner (2013) in terms of ethnobiology. Such an imperative is essential as we see declining numbers of STEM majors and researchers throughout many institutions of higher education in the United States (AAAS 2011; Harrison 2014) and as we see increasing threats to the value of science and scientific knowledge by the Trump administration (Goldman et al. 2017). In addition, what is essential is instilling a morality of stewardship, and this is a main goal with this class. Eugene Anderson, in *Ecologies of the Heart* (1996), has argued that the cornerstone of environmental stewardship is an emotional connection to land and species, as evidenced in many traditional cultures throughout the world. Similarly, Eugene Hunn (2014) describes the linkage (not uncomplicated) between knowing nature, loving nature, and being motivated to act on behalf of nature. As educators in the fields of anthropology, ethnobiology, environmental studies, and related fields, we are uniquely poised to provide such courses for our students. I see this in fact as a moral duty we have to society at large, as ecological literacy is essential to a healthy and functioning society, and perhaps to the survival of our planet.[3]

Assignments in the course include various types of writing, data collection, and engaged learning. This includes weekly writing that is generally process oriented, writing two five-page papers (one on an assigned topic, another on a topic of the students' choosing) with one round of peer-review and another with an extensive review by me, participation in an in-class oral debate (topic and positions assigned), care of a garden plot on campus (not all semesters), a field trip to a local food forest (not all semesters), conducting a pet survey based on a study from 1985, mandala meditation and blogging (see Wolverton, this volume), and gathering data through first-hand empirical observation. The assignment that is central to this chapter is the crow assignment, which has components of reading, observation and data collection, discussion, and writing. Note that this assignment takes only one to two weeks and is one of many assignments. Modifications could be made for this to be a more extensive course project, which I discuss towards the end of the chapter.

Caw Connections: The Assignment

Crows are fairly abundant throughout much of urban and suburban North America (Marzluff and Angell 2005). They are ubiquitous residents on our university campus. Therefore,

3 I admit a slightly selfish goal: as a visiting professor I have had to consider ways to keep my worth relevant to the University. Faculty have been needed to teach Seminars in Scholarly Inquiry (SSi) in part because these courses are a lot of work due to intensity of grading and student contact; the University caps enrollment for SSi courses at 17 students per class. Many faculty, especially those in fields outside of literature, also often feel ill equipped or are otherwise uninterested in teaching about writing. Thus, demand for faculty to teach SSi courses has been great—a fact I was aware of when designing the course.

the opportunity for students to learn more about their immediate surroundings was key when I was designing this assignment. I thought in part about citizen science, and about ways to incorporate common experience with an element of new learning. In addition, the intelligence of crows and ravens (and other birds, especially parrots, for that matter) has always impressed me, and I suspected it might impress students. The result was a successful assignment that is one of nearly a dozen that constitute the core of the class.

Students are assigned two chapters from John Marzluff and Tony Angell's book *In the Company of Crows and Ravens* (2005). The two chapters are "Cultural Connections" (chapter 1) and "Intertwined Ecologies and Mutual Destinies" (chapter 3). The work gives students the opportunity to learn about human and corvid co-evolution, about some cross-cultural ideas about corvids, and about significant corvid behaviors and cognition. The book is based on research by Marzluff and others, written in a popular style of writing with illustrations that help augment student engagement with the text. We discuss both the key points that the authors present and how they present them. If there is time, we may watch clips of the film *A Murder of Crows* (Fleming 2010) or John Marzluff's TED talk (Marzluff 2014). Sometimes I add in points about the popularity of American folk songs (or band names) about crows. The added audio-visual materials are often popular among students, and augment their learning through sensorial experience. Sometimes during the weeks of this assignment, we visit our on-campus Slater Museum of Natural History to look at specimens of corvids, crow and raven eggs (some dating back to the early 1900s), and to discuss adaptation among birds. Some students are particularly eager to touch the corvid specimens, while others are not. We also discuss the development of natural history museums as sites first of collections of "natural oddities" in the eighteenth century to sites for science education and research in the twentieth century and beyond.

The observation part of the assignment is for students to observe crows four times for 15 minutes each time, and to take notes with as much detail as they can (or to video record and then take notes from the video later). They are encouraged to observe at different locations on campus, and at different times of the day, to increase the variability in data. They are not told exactly what aspects to pay attention to, but after the observations we discuss in part what it is that they noticed; some students realize that they paid close attention to sound, others to movement, others to whether people passing by the crows even noticed them. Some students include sketches with their notes (see Figure 2.1), which can give us the chance to talk about visual representation and symbolism, as well as different modalities of documentation. The flexibility in what to focus on in the observation proves to be useful in terms of having students reflect on their own skills of attention and to notice points of connection or divergence from their classmates with these skills. We discuss how these different data points provide different information, and the significance of these different types of information. Using the expert knowledge of crow researcher John Marzluff, we discuss possible explanations for some of the behaviors that students observed.

Figure 2.1. Sketch of crow by Madeline Lampard, a student in Professor Glover's SSi1 *People, Plants, and Animals* course at the University of Puget Sound, Fall 2019.

The last part of the assignment that students complete is composing a 350-word article about crows for the school newspaper. In the article, students must use two effective quotes from the Marzluff and Angell (2005) reading, and three pieces of evidence from their own observations. They are encouraged to write in a creative and compelling voice, to grab the attention of the reader. Students are instructed to entice the reader into observing crows, and into thinking about crows in a way that does not frame them as "pests" but instead in a different (unspecified) way. Students are not required to submit their articles to the newspaper, but are encouraged to do so (thus far, I do not know that any student has actually submitted). Below are some partial excerpts from the articles.

> *It's a Crow's World, We Just Live in It*
> Crows can and should be considered one of the smartest creatures on Earth. They have survived thousands of years accomplishing the task that few animals have, surviving with humans. In my observations of Crows I discovered that we are simply pawns that they use in order to accomplish their goals.

> *Crows: They Don't Need the SAT to Prove Their Intelligence*
> How many crows have you seen today? What have your thoughts been when you have seen them? Was your first impression "ugh another one!"? After reading John M. Marzluff and Tony Angell's book, *In The Company of Crows and Ravens*, you will be late to class because you've been watching the crows hopping in the grass. The book explains the evolution of crows and ravens, their incredible tenacity, and what makes them so perfect to live alongside humans. For our seminar assignment this week we got to watch crows and observe them. I know many would hear this homework and roll their eyes, so what they're just crows they're everywhere nothing makes them special. Just like anything else, crows aren't special until you learn more about them .... Crows are unique to me because they are very capable on their own but choose to take a simpler route by living alongside wolves and/or humans. They finish what we do not. I could go on about crows and how they ought to be respected and even admired but I do not want to be preachy, or even worse, a "bird girl."

What these samples highlight is a creativity grounded in experiential and cognitive learning. They are demonstrations of effectively engaged learning and creation, connecting with crows and encouraging peers to consider doing the same.

Course Outcomes

The results of this brief assignment are impressive. In addition to distinguishing between different types of data (i.e., scholarly, peer-reviewed research and first-hand observations that they themselves conduct), I identify three key results: development of awareness, development of sensitivity, and development of respect. In terms of awareness, students often realize that they never really "noticed" crows on campus, even though they are widespread both on campus and in many urban areas where students are from. This means two things. First is that, in some cases, students literally did not see crows on campus—I even had one student ask where to find crows once, to which I pointed out the window to a crow sitting in a tree outside the classroom. In other cases the meaning is more metaphorical: students would see crows, but could not make much sense of their behaviors and did not think of crows as anything special. In either case, crows become "visible" to students due to the assignment. Some students comment that they had heard the vocalizations of crows before, but did not know where those sounds were coming from (some guessed they were from squirrels, some from other birds), or that they never really even heard the sounds at all before, they simply registered as "noises" in the environment. So, in some cases the assignment also brings voice to the crow.

The second type of growth that this assignment seems to engender is sensitivity. Much of this development is due to being exposed to the expert knowledge of John Marzluff through reading, but there is little doubt in my mind that this learning becomes more deeply experienced through their own observations. This includes sensitivity to the details of crow behavior and what various movements signify. For example, poking beaks in the grass is usually a search for bugs, while wings fluttering outward may be a sign of begging from a young crow. This also includes sensitivity to the fact that crows are social creatures that live in communities. This is a bit more challenging to observe without extended observation (because three crows perched on a branch close to each other, or hopping around the garbage can outside the Student Union Building do not necessarily signify community). Nonetheless, students sometimes do see crows interacting in social ways that they recognize—often in competition for the same resource, for example. And students also often notice crows calling back and forth to each other, which is of course a sign of sociality. There is, quite significantly, a developed sensitivity to the intelligence of crows. This seems to be an especially impressive discovery for students, and is in fact often repeated frequently in student comments about what they learned in the assignment and/or the entire course. Again, much of this comes from the assigned readings and class screenings of films which demonstrate the ability of crows to solve puzzles. Furthermore, a few students over the years say they have witnessed crows dropping things from on high as they are known to do with nuts to crack the shells (an indication of advanced planning capabilities).

Lastly, there is a development of respect for crows that few students seem to even consider before the assignment. Much of this new-found respect does seem linked to realizing the intelligence of crows. Students say that they no longer think of crows as "trash" birds and that they experience a kind of seeing and being seen by crows, eye to eye. They are usually fairly surprised by their own reactions to these feelings of respect and even awe, in some cases. We often talk about this in class as well, and discuss the concept of charismatic megafauna and how that often figures into human ideas about the value of non-human beings. I encourage students to reflect on why their new-found respect for crows is usually linked to crow intelligence, and how identifying commonalities with non-human animals can affect the way we consider them.

These are exactly the results that I had hoped for when designing the course. Below are some quotes from students about this assignment (in response to a prompt about their learning in the class) that capture some of what I explain above.

> My favorite [reading] ... is the crow [reading]. I loved learning about crows' social behaviors and their culture. I found it particularly interesting that crows parallel humans at times, like how they live in the city and then move to more rural areas after maturing.

> After we did the crow assignment in class and learning about how smart they are, whenever I see them on campus, I always stop and take a second to try to figure out what is going on inside of their brains. Because of that assignment, crows are now interesting to me.
>
> The new perspective that stood out to me the most was the realization that animals are just as important to the environment and society as humans. Before I would have thought that humans were the superior living organism, but now I believe that all living things deserve equal respect. An assignment that helped with this realization was the crow assignment. An animal that I used to think was a useless gross species, I discovered is a very intelligent and important part to society.

Recommendations

Possible adjustments to this assignment could include longer or more observation times, additional readings by other crow researchers and/or from cross-cultural perspectives about crows, or using a different writing assignment. I had adjusted the timed observations slightly over a few years (longer observation times of 20 minutes, for example) but have not found much difference in results for student learning. However, a larger sample size of observations could yield important results, and would allow students to understand the benefits of extended observation. Another idea would be to design this as a series of observations to take place over weeks or the entire length of a semester, in part with the hopes of observing variations in crow behavior possibly linked to seasonal changes. A more in-depth look at views about corvids from other cultural perspectives could provide important points of both contrast and connection (e.g., Nelson 1983; Zahara and Hill 2015); some of this is covered in Marzluff and Angell (2005)—for example, many Indigenous cultures in North America have reverence for corvids, in part due to their obvious intelligence (which often manifests in a keen ability to trick humans). An expansion of reading about other cultural viewpoints, especially from Indigenous scholars and naturalists, would increase students' exposure to seeing how much some cultural groups pay close attention to corvids. Related to this, a writing exercise that focused on reading various scholarly sources on corvids (e.g., Goodwin 1976; Massen et al. 2014; Seed et al. 2007; Szipl et al. 2015; Zahara and Hill 2015) and from Indigenous and non-Indigenous perspectives, rather than relying solely on the Marzluff and Angell (2005) book, might help provide other points of view and arguments from grounded research, and could also give an opportunity to discuss different writing styles, as well as Indigenous and non-Indigenous perspectives on corvids and non-human animals.

A concern I sometimes have is whether the observation notes are "real." This is a problem with any assignment that relies on self-reporting. It is usually fairly obvious when notes are embellished or likely fabricated (like once when a student's notes indicated that they heard crows vocalizing in the night) but one could require actual video footage as part of the documentation, if desired.

There should be consideration of "observer bias" if the observation is assigned after the reading. I have opted for this sequence because of a desire for informed seeing on the part of the students; having some notion about what some of the behaviors signify I believe makes for a more fruitful observation experience. However, the assignment could be altered to have observations happen first, and then informed reading happening next. Or, observations could be conducted both before and after the reading assignment, with discussion and/or writing reflection on the changes in understanding students might have.

Many other alterations could be considered in addition to the ones already mentioned. For a course in the humanities, a creative output such as an artistic interpretation of crow life could be fruitful. In the social and natural sciences, a longer research paper is possible, perhaps with an element of contrasting crows with other commonly sighted birds, depending on location.

One of the wonderful features of this assignment is how easily adaptable the assignment can be to various college campuses in North America. Crows are ubiquitous in urban and suburban settings (where many campuses of higher education are located). Crows also live in rural settings but may be less likely to be within observation distance, as experts like Marzluff and Angell (2005) explain that rural crows are a bit more leery of humans and avoid them more frequently than urban and suburban crows do. Most North American campuses also have trees and open lawns, and dining areas with potential "food drops"—all features of landscape that attract crows. Only once in five years since teaching the course has a student said to me that they could not find any crows to observe (I think it was due to the fact that the student was not looking up into the trees, but only on the open lawns… the student did find some to observe after some suggestions by me as to where to look). Thus, it is workable for nearly any North American campus, save those in Hawaiʻi (which isn't in North American anyway).

Concluding Thoughts & Reflection

Many students identify in their observation notes some of the behaviors that Marzluff and Angell (2005) discuss (searching for insects in the grass, or for french fries in the trash), so there is good integration of understanding obtained through reading with first-hand empirical investigation. Some students assign personal names to crows and talk about what they might wear if they were people, exciting their own imaginations about crows as people. These all provide fodder for good class discussions, but they also very importantly allow students to connect to crows in ways that the majority of them never have before. One point in particular seems to be readily appreciated by students, and that is the intelligence of crows. A majority of students comment (either in written form, or orally in class) that they had no idea crows were so smart. At a time in their own lives when they are becoming acutely aware of their own learning and intelligence, to become aware that other creatures, even "mundane" ones such as crows, learn and have impressive mental capacity seems to be a meaningful revelation for most students.

At the end of many of my courses, I give the students time for self-reflection about their learning (or in some cases about their writing). One of the questions I sometimes ask is "What do you think you might remember about this course in five years?" Every semester I get several students commenting on how they will remember the crow assignment, as it caused them to change the way they see crows. In fact, I have had the experience of reconnecting with seniors that took my course as first-years who tell me they remember the assignment! Here are some student reflections in response to my question:

> In five years I expect to remember the crow reading. Crows have always been around in my life but I have never taken the time to notice them until we talked about their behaviors in this course. Ever since then, I have taken the time to notice their behaviors and compare them to other crows I have seen in different areas of Washington State and this country.

> What I expect to remember in 5 years from now is how intelligent crows are. This topic stood out to me because I never expected crows to be even remotely as intelligent as they really are which has completely changed my perspective on these amazing creatures. I never suspected these birds to be highly intelligent or observant and it only makes me more interested in these birds and how humans have viewed them throughout history.

> I expect to remember that crows can remember people's faces.... because I found the crow assignment very fascinating, and I really do not wanna make any winged enemies any time soon.

In summary, this assignment encourages the kind of knowing and caring that is necessary to develop a sense of environmental stewardship, a core goal of the class. Engaging in the intellectual work of careful reading of scholarship, gathering first-hand empirical evidence, and then analyzing and synthesizing this new-found knowledge with data is coupled with a creative production of writing, grounded in real-world evidence. This is successful, engaged learning. Whether this assignment truly engenders love (cf. Hunn 2014) of crows, or nature more broadly, is not clear; it does seem to engender respect, which is a step in the right direction. It is naïve to think that such a brief assignment would necessarily be enough to engender true love, and indeed one wonders if a one-semester course would even be enough time to develop such a deep commitment, which takes time and dedication. But I have been pleased with the results of this modest assignment; respect is a powerful starting place for life-long commitment.

References Cited

AAAS (American Association for the Advancement of Science). 2011. Vision and Change in Undergraduate Biology Education: A Call to Action [web page]. AAAS, Washington, DC. URL: http://visionandchange.org/finalreport. Accessed on July 13, 2019.

Anderson, E.N. 1996. *Ecologies of the Heart: Emotion, Belief, and the Environment.* Oxford University Press, Oxford.

Fleming, S. dir. 2010. A Murder of Crows [Film]. Documentary. 60 min. Produced by WNET and Susan Fleming. Public Broadcasting Service (PBS), Arlington, VI.

Goldman, G.T., E. Berman, M. Halpern, C. Johnson, Y. Kothari, G. Reed, and A.A. Rosenberg. 2017. Ensuring Scientific Integrity in the Age of Trump. *Science* 355(6326):696–698.

Goodwin, D. 1976. *The World of Crows.*The British Museum and Comstock Publishing Associates, London.

Harrison, P. 2014. Carrying Plant Knowledge Forward in the USA. In *Innovative Strategies for Teaching in the Plant Sciences*, edited by Cassandra L. Quave, pp. 3–20. Springer, New York.

Holm, B., V.C. Carter, and T. Woodin. 2010. Vision and Change in Biology Undergraduate Education: Vision and Change from the Funding Front. *Biochemistry and Molecular Biology Education* 39:87–90.

Hunn, E. 2014. To Know Them is to Love Them. *Ethnobiology Letters* 5:146–150.

Marzluff, J. 2014. Crows, Smarter Than You Think [web page]. TEDxRainier talk. URL: https://www.youtube.com/watch?v=0fiAoqwsc9g. Accessed on December 28, 2018.

Marzluff, J., and T. Angell. 2005. *In the Company of Crows and Ravens.* Yale University Press, New Haven, CT.

Marzluff, J., and T. Angell. 2012. *Gifts of the Crow: How Perception, Emotion, and Thought Allow Smart Birds to Behave Like Humans.* Free Press, Glencoe, IL.

Massen, J.J.M., G. Szipl, M. Spreafico, and T. Bugnyar. 2014. Ravens Intervene in Others' Bonding Attempts. *Current Biology* 24:2733–2736.

Nelson, R.K. 1983. *Make Prayers to the Raven: A Koyukon View of the Northern Forest.* University of Chicago Press, Chicago.

Seed, A.M., N.S. Clayton, and N.J. Emery. 2007. Postconflict Third-Party Affiliation in Rooks, *Corvus frugilegus. Current Biology* 17:152–158.

Szipl, G., M. Boeckle, C.A.F. Wascher, M. Spreafico, and T. Bugnyar. 2015. With Whom to Dine? Ravens' Responses to Food-associated Calls Depend on Individual Characteristics of the Caller. *Animal Behavior* 99:33–42.

Wagner, G. 2013. *Vision and Change for Undergraduate Ethnobiology Education in the U.S.A.* Botanical Research Institute of Texas Press, Fort Worth, TX.

Zahara, A.R.D., and M.J. Hird. 2015. Raven, Dog, Human: Inhuman Colonialism and Unsettling Cosmologies. *Environmental Humanities* 7:169–190.

CHAPTER 3.

Working with Indigenous Elders in Biology and Chemistry University Courses

Fidji Gendron[1], Vincent Ziffle[1], and Nametha Srikanthan[2]

Chapter Summary

This chapter describes the participation of Indigenous Elders in undergraduate biology and chemistry courses and examines students' perception of the role of Indigenous Elders in these courses. The percentage of Indigenous people with high school or post-secondary education has increased in the last ten years in Canada but is still significantly lower than the rest of the population. Indigenous students often feel a disconnection between Indigenous science and Western science and the inclusion of Traditional Knowledge by Elders in the classroom can consolidate these two ways of learning. Hands-on laboratory activities (e.g., hide tanning, hand lotion with native plants, and porcupine quill bookmarks) were developed under the guidance of an Elder. Students ($n = 258$) were asked to complete a survey after they performed these traditional activities in the biology laboratory and participated in discussions about medicinal plants in their chemistry classroom with Elders. Students agreed or strongly agreed that Elders help them understand the natural world better, are effective communicators, expand their scientific knowledge, create a more inviting classroom, are culturally relevant scientific role models, and add meaning to what they are learning in their biology or chemistry courses. Thematic analysis revealed that students appreciate the presence of Elders, the traditional ways of performing the activities, and the hands-on approach. Students suggested to increase the time spent with Elders and to have more hands-on activities. The

1 Indigenous Knowledge and Science Program, First Nations University of Canada, 1 First Nations Way, Regina, Saskatchewan, Canada, S4S 7K2

2 College of Nursing, University of Saskatchewan.

activities presented in this chapter represent a starting point for educators who are interested in working with Elders to weave local knowledge and stories into university science courses.

Introduction

The participation of Indigenous Elders in science courses can transform education and provide learning opportunities for all students. Their presence can help bridge the education achievement gap between Indigenous people and the rest of the population. In Canada, the legal term "Aboriginal peoples" includes the Indian, Inuit, and Métis peoples who represent 4.9% of the national population. Most commonly referred to as "Indigenous," the Canadian Indigenous population is relatively young (the average age is 32.1 years compared to 40.9 years for the Canadian non-Indigenous population) with the overall Indigenous population exhibiting four times the growth rate of the non-Indigenous population from 2006 to 2016 (Statistics Canada 2017a). In 2016, 86.3% of adults in Canada had completed high school or obtained an equivalency certificate, with 28.5% of them having completed a Bachelor's degree or higher post-secondary degree. While lower overall, the percentage of Indigenous people with high school or post-secondary education has increased from 2006 to 2016 and is now at almost 70%. The percentage of Indigenous people who have completed university degrees was lower than the rest of the population in 2016, but this number had increased from 7.7% to 10.9% since 1996 (Statistics Canada 2017b).

One obstacle facing Indigenous people in the obtainment of a Bachelor's degree is that they often live considerable distances from cities where universities are located. Thus, Indigenous people living off reserves are more likely to have completed a Bachelor's degree or higher (11.4%) compared with 5.4% of those living on reserves (Statistics Canada 2017b). This trend continues in academia where there is markedly lower representation of Indigenous educators and researchers in STEM (Science, Technology, Engineering, and Mathematics) fields. As such, it has been recommended that efforts should be made to promote and provide long-term support for Indigenous science, technology, engineering, and math research, with the goal of increasing the prevalence of research and training by and with Indigenous researchers and communities in Canada (Advisory Panel on Federal Support for Fundamental Science 2017).

The terms "Indigenous science" and "Indigenous knowledge" are used interchangeably in this chapter, as are "Indigenous Elders" and, more succinctly, "Elders." According to Ogawa (1995:588), Indigenous science is defined as "a culture-dependent collective rational perceiving of reality, where 'collective' means held in sufficient similar form by many persons to allow effective communication, but independent of any particular mind or set of minds." It is a strongly local, place-based community phenomenon (Aikenhead and Elliott 2010). Some authors argue that the term "Indigenous science" should be used instead of "Indigenous knowledge" as it places it on equal footing to Western science (Snively and Williams 2016).

Other authors use Indigenous science as a synonymous term for ways of knowing, traditional ecological knowledge, and Indigenous knowledge (Aikenhead and Elliott 2010), which we adopt in this chapter. It is also acknowledged that Indigenous peoples are tremendously diverse, and so too are their knowledges (Kapyrka and Dockstator 2012).

Indigenous students taking science courses often feel a disconnection between Indigenous science and Western science, where the latter can be perceived as exclusionary and hostile to traditional ways of knowing (Kimmerer 2002). As such, inclusion of Indigenous science in Western science curricula may make such courses more welcoming for Indigenous students (Kimmerer 2002). Indigenous students appreciate when their ways of knowing are acknowledged in the science classroom (Aikenhead and Elliott 2010) and come to a deeper understanding of connections between Indigenous science and Western science (Baker et al. 2016), leading to greater success in courses overall. Learning environments that recognize the embeddings of culture in everyday practices should support student learning and motivation (Bang and Medin 2010). The inclusion of Indigenous science also has the benefit of expanding the horizons of students from the dominant culture (Kimmerer 2002) and can nurture a richer understanding of the physical world (Aikenhead and Elliott 2010). In Canada, the inclusion of Indigenous knowledge in education has been emphasized with calls to action issued by the Truth and Reconciliation Commission mandated to examine the abuse inflicted on Indigenous peoples through the Indian residential school system. The Commission developed guiding principles of truth and reconciliation, one of them being "Supporting Indigenous peoples' cultural revitalization and integrating Indigenous knowledge systems, oral histories, laws, protocols, and connections to the land into the reconciliation process are essential," (Truth and Reconciliation Commission of Canada 2015).

Elders can be the bridge between Indigenous science and Western science for students who feel a disconnection between their worldview and what they learn in school. Elders have earned the respect of their community and possess a deep spirituality that is ingrained in their lives and teachings. Elders are committed to sharing their traditional knowledge and providing guidance while respecting their principles and values, and they show their knowledge by providing examples (Joseph 2012). In post-secondary education, Elders in the classroom strengthen the students' ability to understand the relevance of the curriculum by relating it to their cultural background. They also work with instructors to find new approaches to incorporate Indigenous ways of learning into the instructor's teaching methods (Anonson et al. 2014). The inclusion of Indigenous science in the curriculum has the potential to nuance Eurocentric teachings, which often discredit other knowledge bases and values and elevate one source of knowledge over other worldviews (Battiste 1998).

Natural sciences, such as biology and chemistry, are disciplines that can benefit from braiding of Indigenous and Western science. Biology examines phenomena related to living organisms, while chemistry covers elemental and molecular properties and chemical interactions within these biological organisms. Both disciplines strive to lend levels of understanding to organisms living in the natural world. In parallel, traditional ways of knowing are characterized by observation, participation, and experiential learning, and a science classroom or laboratory is a physical space where such activities can take place.

This chapter describes the participation of Elders in biology and chemistry courses and students' perception of the role of Elders in these courses. The activities performed in a biology laboratory were selected to represent traditional practices to support student learning and motivation (Bang and Medin 2010). It is hoped that this chapter will spark interest and reflection regarding potential avenues of how Indigenous and Western science can be braided into the science curriculum to improve learning outcomes for all students. The goals of the study are: 1) to develop hands-on laboratory activities under the mentorship of an Elder, and 2) to examine how students perceived the participation of Elders and the inclusion of Indigenous science in these activities.

Materials and Methods

Participants

Two hundred and fifty-eight students ranging from grades 6–12 to the undergraduate university level participated in this study (Table 3.1) by attending laboratory activities or class presentations at the First Nations University of Canada, Regina, Saskatchewan from January 2016 to April 2017. University students were registered in Biology 100 (*From Cells to Organisms*), Biology 101 (*Organisms in their Environment*), Chemistry 100 (*Introductory Chemistry*), Chemistry 104 (*General Chemistry I*), and Chemistry 140 (*Organic Chemistry I*) taught by Fidji Gendron (FG) and Vincent Ziffle (VZ). Elementary and high school participants were students from various Indigenous community schools near Regina. The First Nations University of Canada has been offering science workshops for the past four years, and workshop advertisements were sent to Indigenous community schools within and surrounding Regina. Teachers who were interested in the program contacted FG to be a part of the laboratory

Table 3.1. Demographic and characteristics of participants (N = 258).

Gender	%
Female	60
Male	39
Ancestry	
First Nations	26
Inuit	0
Métis	5
Other	67
Education	
School	11
University	86
First experience with an Aboriginal Elder	
Yes	33
No	61
This is the first time I have learned about Indigenous knowledge	
Yes	24
No	75

Note: Percentages may not add up to 100% because participants could leave the question blank.

activities. All participants were required to sign a consent form and youth participants were required to have their parent/guardian sign the form received by their teacher. The project was conducted with approval from the Research Ethics Board at the University of Regina.

Description of the Activities

Three biology laboratory activities are described in Appendices 1–3 and were performed under the guidance of Elder Betty McKenna with different groups of students. Elder Betty is from the Anishinaabae Nation from the Shoal River Band in Manitoba and acquired her knowledge from her grandmothers and other Elders. The nature of the laboratory activities was decided by Elder Betty and FG. The first activity consisted of hide tanning. Students learned how to clean the hide, built a frame to stretch it, and removed the flesh and fatty tissues. Traditional (such as sharpened elk rib bones) and modern scraping tools were used to remove the flesh and fat from the hide. It took several days to perform all the steps, so different groups of university and school students were involved in hide tanning. In the second activity, students made a hand lotion with the native prairie sage plant. The leaves were ground using traditional tools, such as grinding stones, and a coffee grinder and left to infuse in oil for two hours. Elder Betty McKenna shared the knowledge that bear grease was traditionally used for the lotion, noting how people would follow bears to get to a bee's hive to get the wax. At the end of the activity, each student went home with a sample of the lotion. In the third activity, students learned how to sew with porcupine quills (obtained from roadkill). While students were sewing, Elder Betty McKenna explained how people would harvest the quills without harming the animal by throwing a blanket over a porcupine and retrieving the quills embedded in the blanket. In the past, all these activities would have been done outdoors so the Elder compared how it used to be done and what changes needed to be done in a laboratory setting. A more complete description of these activities is provided in Appendices 1–3. While students performed the activities, the Elder shared stories and discussed the importance of these activities in her culture. University biology students and elementary and high school students participated in these workshops. These activities were done during regular laboratory sessions for biology students.

Harold Lavallee was invited during the chemistry lectures to discuss the importance of medicinal plants. Harold is a Traditional Knowledge holder and self-described Medicine Man from the Piapot First Nation in Saskatchewan. He prefers to omit the Elder title. Harold discussed the characteristics of medicinal plants as well as their cultural significance, often by telling stories of his upbringing and how he was impacted by the plants and their use in times of need. Harold brought samples of each of the medicinal plants he normally collects to the classroom discussion and explained the challenges related to harvesting rare plants including the distances traveled, protocols followed during harvest, and the impact of the current environment on these traditional activities. Each session was followed by a discussion period with the students, often leading to long, heartfelt discussions about the significance of Indigenous medicinal plant wisdom and traditional ecological knowledge.

These projects honored Indigenous culture and followed proper traditional protocols. Before each activity and presentation, a student volunteered to present tobacco to the Traditional knowledge holder in the form of a tobacco tie and cigarettes, depending on his or her preferences. The Elder and Medicine Man discussed the importance of tobacco offering in their culture before their activity or presentation. Elder Betty and Harold were each offered an honorarium.

Survey

At the end of each activity or presentation, students were asked to complete a survey. The survey consisted of three parts, taking approximately 15 minutes to complete. In the first part, students answered background questions (e.g., their gender, ethnicity, education level) and shared their familiarity with Indigenous knowledge. The second part consisted of quantitative questions; students used a four-point Likert scale to rate if they agreed with statements such as "The knowledge shared by the Elder helps me understand what is covered in my Biology/Chemistry course" and "I believe Elders are effective educators". The third part consisted of four qualitative, open-ended questions: "What was the part of this activity you appreciated most?"; "What part of the activity did you like the least?"; "Did you learn something new and interesting today? And if so, can you tell us what it was?"; and "How can we improve this activity?" Qualitative answers were analysed using thematic analysis where major themes were identified inductively. Participant phrasing was retained.

Results

The demographics and characteristics of the 258 participants are presented in Table 3.1. The majority of the participants attended university (86%) and was mostly comprised of female students (60%). Thirty-one percent of the participants were Indigenous, while 67% were non-Indigenous. This was the first experience with an Indigenous Elder for one-third of the students but 75% of them had learned about Indigenous knowledge before. Only one elementary or high school student did not have prior experience with an Elder.

Almost three-quarters of the students strongly agreed that Elders are a valuable source of Indigenous knowledge (71.7%; Table 3.2). The impact of Elders was important as more than 80% of the participants partially agreed or agreed that Elders helped them understand the natural world better, are effective educators, expanded their scientific knowledge, created a more inviting classroom, are culturally relevant scientific role models, made them appreciate their course more because of the activity, and added meaning to what they learned in their biology or chemistry course. Most of participants partially agreed or strongly agreed (41.7%, 53.0%, respectively) that they appreciated the merging of Western science and Indigenous knowledge. On the other hand, 7% of the participants strongly disagreed that

Table 3.2. Percentage of participants who answered questions on the role of Elders in science.

	I strongly disagree	I partially disagree	I partially agree	I strongly agree
I would like to learn more Indigenous knowledge	1.2	3.9	43.4	50.0
Elders are a valuable source of Indigenous knowledge	0.8	1.2	25.2	71.7
Because of this activity, I am now more interested in science	1.2	19.4	53.1	26.0
I find it easier to learn about science when I do activities with Elders	3.1	20.2	50.0	23.6
I believe Elders help me understand the natural world better	1.6	6.2	36.9	53.1
I believe Elders are effective educators	1.2	1.2	38.0	57.0
I believe Elders expand my scientific knowledge	2.7	14.0	57.4	24.0
I would like to have more courses/activities with Elders	2.3	6.2	47.7	41.5
The presence of Elders creates a more inviting classroom	0.8	7.0	45.8	44.6
Elders are culturally relevant scientific role models	2.3	8.5	52.7	33.7
The knowledge shared by the Elder helps me understand what is covered in my Biology/Chemistry course[a]	7.0	18.7	55.6	16.1
I appreciate my course more because of the activity with the Elder[a]	1.3	11.7	53.9	30.9
I believe Elders add meaning to what I am learning in my Biology/Chemistry course[a]	2.2	12.2	54.4	28.7
I appreciate the merging of Western science and Indigenous knowledge[a]	0.0	3.5	41.7	53.0

Note: Percentages may not add up to 100% because participants could leave the question blank.
[a] Questions answered by university students only.

the knowledge shared by Elders helped them understand what is covered in the biology or chemistry course.

When asked about the part of the activity they appreciated the most, the major themes discussed by participants in the open-ended questions of the survey centered around the Elders (their presence, teachings, and stories), Indigenous ways of doing (how it was done in the past and how it is still being done), the hands-on approach to the activity (grinding plant material, scraping flesh on elk hide, sewing porcupine quills), and plant uses. One participant illustrated the importance of the Elders by sharing "The fact that an elder was brought into class using natural medicine/Indian medicine to set examples was very valuable and appreciated," while another enjoyed "Getting background knowledge as to how Indigenous cultures used their resources in the past and was fascinated to hear how they still use plants as resources today."

When asked about the part of the activity they liked the least, a few themes emerged in the open-ended questions of the survey: 1) the elk hide tanning experiment (several students did not like the smell and the scraping and holding of the hide); 2) the organization of the laboratory or lecture (students wanted to work in smaller groups in the laboratory and they wanted to touch and smell the plant samples during the lectures); 3) students did not understand the relationship between the course and the Indigenous teachings; and 4) the activity or workshop was too short. Regarding these last three themes, one participant shared, "I didn't like how I could not see all the time, in particular the demonstration of the sewing technique"; another shared, "I don't know how it relates to our course"; and another stated, "I hate that our time with our Elder was limited."

In the open-ended questions of the survey, we asked participants if they had learned something new and interesting. Three novel themes emerged. First, participants learned that plants have different purposes, such as healing and hand lotion that are still in use today. One participant shared that "I learned where some of the medicines can be found or what they can be used for." The second novel theme centered around traditional knowledge, stories, and protocol and one participant stated "I learned some of the reason that sage is sacred" while another added "I learned about indigenous culture and how to respect nature!" Ways that Indigenous peoples utilized plants and animals in the past was the third novel theme and one participant shared "That in the past they didn't use needles; instead they used bone. The thread was from the back of a moose. I learned that the hand shake is given with the hand closest to the heart."

Participants suggested some recommendations on how to improve the activities in the open-ended questions of the survey. Their main comments were to increase the amount of time spent with the Elder, to pass samples around or provide pictures during presentations in the classroom, to have hands-on activities instead of presentations, to have students work in small groups during hands-on activities in the laboratory, and to provide more empirical data. For example, one participant asked "Would it be possible for Harold to hold workshops where he takes us into the park where we learn to identify medicinal plants and how to collect them respectfully?"

Discussion

According to Battiste (2002), Elders and knowledge keepers should be viewed as living educational treasures because of their knowledge and teachings. One way to remain true to the ways of knowing and knowledge systems of Indigenous peoples is to encourage collaboration among mainstream educators with recognized community Elders who can share their teachings (Bartlett et al. 2012). This is the approach that was used in the present study, and the presence of an Elder in the laboratory and classroom was perceived as positive by the vast majority of the participants.

Culturally based stories about plants and animals that convey knowledge about nature, such as the ones shared in this chapter, are important ways of learning for Indigenous students to engage with science (Bang and Medin 2010). In a study focusing on Indigenous scientists, it was revealed that stories heard during youth that explain the world (e.g., animal migration, medicinal plants) had a profound impact on their desire to become scientists (Quinn 2011).

It follows, then, that for developing post-secondary Indigenous environmental education programs in Canada, Simpson (2002) recommends that Elders be viewed as experts who can offer both guidance to students and instructors and not only as extras or guest speakers. The inclusion of storytelling and Elders in the science classroom promotes participation by and success of university Indigenous students (Hogue 2012). Sharing of Indigenous ways of knowing by Elders validates Indigenous science and likely motivates Indigenous students to connect or reconnect with their culture (Hatcher et al. 2009). These activities expose students to the art of oral traditions, as opposed to reading the information in textbooks. Oral traditions are a key component of Indigenous science where stories and metaphors are used to organize and classify information that is then handed down through generations and to other groups. These stories are viewed as hypotheses or possible explanations of phenomena that can be tested rather than as unreliable anecdotes as is often portrayed in Western science.

A more inviting classroom benefits non-Indigenous participants and Indigenous students who are likely to feel that their culture is welcomed and valued in the classroom. This sense of cultural safety likely increases participants' comfort within Western scientific disciplines traditionally characterized as authoritative, canonical, and theoretical (Carm 2014). Cultural safety is recognized in the public health community as actions that increase, value, or empower the cultural identity and wellbeing of an individual (National Aboriginal Health Organization 2006). In an educational environment, this means that each student's knowledge and reality is recognized as valid and valuable. The challenge for science professors is to teach the science curriculum without altering Indigenous students' values and beliefs and to help them understand new concepts and to explore the differences and similarities between Indigenous science and Western science concepts (Snively and Corsiglia 2001). In a study working with Native American university students in Montana and Arizona, it was reported that cultural activities such as traditional practices enhance feelings of belonging and foster positive academic experiences in science, technology, engineering, and math disciplines (Smith et al. 2014). As such, the inclusion of Indigenous activities, stories, and knowledge in science courses does much to facilitate Indigenous students' academic success. This is true not only for Canadian Indigenous students but for students of non-Western cultures in general. For example, by utilizing the Taiwanese Indigenous students' worldviews as a foundation and combining Indigenous knowledge and Western science in school, teachers were able to motivate interest and to support higher academic achievement (Lee et al. 2012).

An important theme discussed by participants was their interest in hands-on activities. Battiste (2002) writes that Indigenous learning is characterized by experiential knowledge where the person learns by observing, listening, and participating with minimal intervention and instruction, which was the type of learning environment fostered in the biology laboratory exercises in this study. Although participants enjoyed these hands-on activities, they recommended a few ways to improve them such as working in smaller groups and having the possibility of interacting with plants and other items.

In this project, we have developed and organized activities to respect different ways of knowing and to orient our curriculum so that students can benefit from the strengths of both Indigenous science and Western science. To guide us through our learning journey, we invited Elders into our laboratory and classroom to share their knowledge with our students. We have also developed booklets to help other instructors who would like to bring Indigenous science into their curriculum in a hands-on way. This notion of openness towards different ways of learning is referred to as two-eyed seeing and refers to learning to see from one eye with the strengths of Indigenous ways of knowing and from the other eye with the strengths of Western ways of knowing, thus ultimately using both eyes together (Hatcher et al. 2009). In environmental education, this is also referred to as the two-worlds approach in which the Indigenous and Western worldviews are not merged together but are acknowledged and included equitably in the curriculum (Kapyrka and Dockstator 2012). It is believed that understanding the difference between these two ways of knowing increases learning and mutual understanding (Kapyrka and Dockstator 2012).

Although the long-term goal of this project is to offer both Indigenous science and Western science to our science students, several challenges must be overcome to achieve this. While our home university is very supportive of weaving Indigenous science into our science curriculum, the introductory biology and chemistry courses require components that need to be taught within a limited timeframe, which restricts the lengths and numbers of contact times with the Elders. Because these activities and talks were also new to us, we sometimes struggled to find ways to present the associated science components during our lectures, but we feel this can be improved when we next offer the courses. Despite these limitations, these activities paved the way to fruitful relationships with Elders, and we will continue to work with them to facilitate the educational process for our students and to enhance the appreciation of Indigenous science. Working on the land is a critical factor in Indigenous ways of teaching and learning. Field work was not the subject of the current study, but a potential avenue towards land-based learning would be to find ways to bring Indigenous and Western science together beyond the laboratory on the land and/or in a Traditional Campus space.

In this chapter, we hope that we have planted seeds of thought and reflection for science professors who desire to braid Indigenous science into their courses. Our experiences serve as a source of ideas for others, and there is no one-size-fits-all pedagogy for what Indigenous science should look like in a university setting. Traditional knowledge is not a book-based process of learning, and is strongly placed-based. As such, the activities presented in this chapter should be seen as a starting point for educators to reach out to recognized community Elders and knowledge keepers to weave local knowledge and stories with Western science components. As stated by Bartlett et al. (2012), Indigenous knowledge is living knowledge. We believe that successful teaching practices will be the ones that can efficiently bridge home, community, and cultural knowledge with university knowledge.

Acknowledgements

We are grateful to Elder Betty McKenna and Harold Lavallee who guided the activities and the presentations during this project. They are a great source of inspiration for us. We thank the participants in this project. We also thank Leanne Stricker, Trisha Campbell, and Jody Bellegarde who helped us during the Biology activities and the writing of the booklets. This work was supported by the University of Regina President's Teaching and Learning Scholars Grant Program.

References Cited

Advisory Panel on Federal Support for Fundamental Science. 2017. Investing in Canada's Future: Strengthening the Foundations of Canadian Research [web page]. URL: http://www.sciencereview.ca/eic/site/059.nsf/vwapj/ScienceReview_April2017.pdf/$file/ScienceReview_April2017.pdf. Accessed on March 21, 2019.

Aikenhead, G.S., and D. Elliott. 2010. An Emerging Decolonizing Science Education in Canada. *Canadian Journal of Science, Mathematics and Technology Education* 10:321–338.

Anonson, J., S. Huard, T. Kristoff, V. Clarke-Arnault, V. Wilson, and M.E. Walker. 2014. The Roles of Elders in Post-Secondary Educational Institutes. *The Canadian Journal of Native Studies* 34:1–18.

Baker, J., M. Whitstone, S. Yu, T. Roadhouse, and N. Barr. 2016. The Impact of Indigenous Knowledge in Science Education on Urban Aboriginal Students' Engagement and Attitudes Toward Science: A Pilot Study. Urban Aboriginal Knowledge Network, Prairie Research Centre [web page]. URL: https://uakn.org/wp-content/uploads/2015/05/PRC-UAKN-Final-Paper_Impact-of-Indigenous-Knowledge-in-Science-Education-on-Urban-Aboriginal-Students_2016.pdf. Accessed on March 21, 2019.

Bang, M., and D. Medin. 2010. Cultural Processes in Science Education: Supporting the Navigation of Multiple Epistemologies. *Science Education* 94:1008–1026.

Bartlett, C., M. Marshall, and A. Marshall. 2012. Two-Eyed Seeing and Other Lessons Learned Within a Co-Learning Journey of Bringing Together Indigenous and Mainstream Knowledges and Ways of Knowing. *Journal of Environmental Studies and Sciences* 2:331–340.

Battiste, M. 1998. Enabling the Autumn Seed: Toward a Decolonized Approach to Aboriginal Knowledge, Language, and Education. *Canadian Journal of Native Education* 22:16–27.

Battiste, M. 2002. Indigenous Knowledge and Pedagogy in First Nations Education: A Literature Review with Recommendations. National Working Group on Education and the Minister of Indian Affairs, Indian and Northern Affairs Canada, Ottawa, Ontario [web page]. URL: https://www.afn.ca/uploads/files/education/24._2002_oct_marie_battiste_indigenousknowledgeandpedagogy_lit_review_for_min_working_group.pdf. Accessed on March 21, 2019.

Carm, E. 2014. Inclusion of Indigenous Knowledge System (IKS)—a Precondition for Sustainable Development and an Integral Part of Environmental Studies. *Journal of Education and Research* 4:58–76.

Hatcher, A., C. Bartlett, A. Marshall, and M. Marshall. 2009. Two-Eyed Seeing in the Classroom Environment: Concepts, Approaches, and Challenges. *Canadian Journal of Science, Mathematics and Technology Education* 9:141–153.

Hogue, M.M. 2012. Interconnecting Aboriginal and Western Paradigms in Post-Secondary Science Education: An Action Research Approach. *Journal of the Canadian Association for Curriculum Studies* 10:77–114.

Joseph, B. 2012. Aboriginal Elder Definition. Indigenous Corporate Training, Inc. [web page]. URL: https://www.ictinc.ca/blog/aboriginal-elder-definition. Accessed on March 21, 2019.

Kapyrka, J., and M. Dockstator. 2012. Indigenous Knowledges and Western Knowledges in Environmental Education: Acknowledging the Tensions for the Benefits of a 'Two-Worlds' Approach. *Canadian Journal of Environmental Education* 17:97–112.

Kimmerer, R.W. 2002. Weaving Traditional Ecological Knowledge into Biological Education: a Call to Action. *BioScience* 52:432–438.

Lee, H., C.F. Yen, and G.S Aikenhead. 2012. Indigenous Elementary Students' Science Instruction in Taiwan: Indigenous Knowledge and Western Science. *Research in Science Education* 42:1183–1199.

National Aboriginal Health Organization. 2009. Cultural Competency and Safety in First Nations, Inuit and Métis Health Care [web page]. URL: http://rapworkers.com/wp-content/uploads/2017/07/5.CulturalCompetency_Safety-in-First-Nations_Innuite_Metis-health-care.pdf. Accessed on March 21, 2019.

Ogawa, M. 1995. Science Education in a Multiscience Perspective. *Science Education* 79:583–593.

Quinn, R. 2011. The Sacred Tree: Weaving the Stories of Western Science and Aboriginal Ways of Knowing. Master's Thesis. School of Communication & Culture, Royal Roads University, Victoria, British Columbia, Canada [web page]. URL: https://www.collectionscanada.gc.ca/obj/thesescanada/vol2/002/MR76008.PDF. Accessed on March 21, 2019.

Simpson, L. 2002. Indigenous Environmental Education for Cultural Survival. *Canadian Journal of Environmental Education* 7:13–25.

Smith, J.L., E. Cech, A. Metz, M. Huntoon, and C. Moyer. 2014. Giving Back or Giving Up: Native American Student Experiences in Science and Engineering. *Cultural Diversity and Ethnic Minority Psychology* 20:413–429.

Snively, G., and J. Corsiglia. 2001. Discovering Indigenous Science: Implications for Science Education. *Science Education* 85:6–34

Snively, G., and W.L. Williams. 2016. Braiding Indigenous Science with Western Science. In *Knowing Home: Braiding Indigenous Science with Western Science*, edited by G. Snively and W.L. Williams, ch. 1. University of Victoria Press, British Columbia, Canada.

Statistics Canada. 2017a. Aboriginal Peoples in Canada: Key Results from the 2016 Census [web page]. URL: http://www.statcan.gc.ca/daily-quotidien/171025/dq171025a-eng.htm. Accessed on March 21, 2019.

Statistics Canada. 2017b. Education in Canada: Key Results from the 2016 Census [web page]. URL: http://www.statcan.gc.ca/daily-quotidien/171129/dq171129a-eng.htm. Accessed on March 21, 2019.

Truth and Reconciliation Commission of Canada. 2015. What We Have Learned: Principles of Truth and Reconciliation [web page]. URL: http://www.trc.ca/websites/trcinstitution/File/2015/Findings/Principles_2015_05_31_web_o.pdf. Accessed on March 21, 2019.

Chapter 4.

Experiential Herbal Medicine Lesson

Amanda Thiel[1]

Chapter Summary

"Welcome to the garden," I say, borrowing an opening line from James Green's (2011:341) *Herbal Medicine Maker's Handbook* to begin my *Local Medicinal Plant Identification and Herbal Medicine-Making* workshops. Standing outside in the elements, surrounded by the living world—I greet those gathered by stating the obvious as an easy act of friendship.

What happens next is the experiential, outdoor, medical ethnobotany workshop, which I have presented to groups of local university students almost annually over the last decade. The workshop is an introduction to local plants and their medicinal uses in the Western U.S. herbal tradition, and instructs in the methods and rationales of preparing an herbal extract from plants harvested throughout the educational plant walk.

In this chapter, I discuss what works and what does not with outdoor instruction with a large group, and how I have refined this particular presentation into an effective outdoor educational opportunity. I discuss lesson content, student feedback, and make suggestions for adapting this workshop to other courses and institutions. I also discuss pedagogical elements of the lesson. I address the importance of developing a "community of inquiry" through cognitive, social, and instructor presence, which creates a dynamic learning experience in which students engage with content, each other, and the instructor in constructive ways (Garrison et al. 2010; Splitter and Sharp 1995). I also discuss how the lesson contributes to an environmental ethic—shaped by the "direct encounter" with nature and the corresponding ethical implications in the way we talk about and interact with nature—that has potential lifelong impacts outside of the classroom (Rozzi 1999; Rozzi et al. 2006:43).

So let's head to the garden. It's where we all want to be anyway.

---

1 Washington State University, Department of Anthropology, College of Arts and Sciences

The Institution and Course

Today, the *Local Medicinal Plant Identification and Herbal Medicine-Making* workshop is an invited guest lecture for an undergraduate Ethnobotany course at Fairhaven College, Western Washington University, in Bellingham, Washington. It began as a lesson for an Applied Human Ecology course, and I presented it there twice. It has since become a mainstay in the Ethnobotany course, wherein I have presented it six times. With an average of 22 students in each class, I have presented it to approximately 175 students over the last ten years.

Western Washington University ("Western") is a public institution with 16,000 students (13,000 when I began these lessons ten years ago) in Bellingham, Washington, about 145 km (90 miles) north of Seattle and 80 km (50 miles) south of Vancouver, British Columbia. Western's location is in the maritime climate of the Pacific Northwest, nestled between the Cascade Mountains and the Salish Sea. Its focus on innovative teaching allows for many diverse programs that provide unique learning opportunities for students. These opportunities include an environmental education program that makes use of place-based learning in the North Cascades Mountains, and several interdisciplinary and self-designed undergraduate degree programs.

Fairhaven College is one such interdisciplinary, liberal arts, undergraduate degree program within Western Washington University, that allows students the option to self-design their concentration (or major). With an enrollment of approximately 400 students, class sizes are small (usually 10–15 students) and advisor-student mentorship is strong. Students and faculty use narrative evaluations as opposed to grades to assess student work. There is considerable emphasis on independent and experiential learning opportunities. While diversity is increasing, Fairhaven College, like the larger university, is predominantly attended by white, middle-class students.

As I mentioned, I have presented the *Local Medicinal Plant Identification and Herbal Medicine-Making* workshop to two courses at Fairhaven College. One was Fairhaven's 200-level Applied Human Ecology course, which is no longer offered. This course was organized and co-taught by upper-level undergraduate students under the mentorship of Dr. Gary Bornzin. It covered a range of topics relating to people's relationship with the environment with an aim of promoting sustainability. The other course is a 300-level Ethnobotany course, taught by Dr. John Tuxill. It discusses theoretical and practical material about the interrelationship of people and plants. The course spends roughly a week on each topic, from botanical nomenclature and local plant identification to plant foods and more, and, naturally, my lesson fit into the week on medicinal plants. Through the course structure, students gain exposure to the topic of medicinal plants through readings, lecture, and discussion, before engaging in the practical, hands-on, outdoor portion of the topic which happens through attendance at my 2-hour workshop. Both the Applied Human Ecology and Ethnobotany courses satisfy one of the science requirements at Fairhaven College, although many students choose them as electives. The courses have limited space available to non-Fairhaven students, so some students are in other degree programs at Western.

My lesson is an outdoor workshop that makes use of "the Outback," Western Washington University's on-campus, 2-ha (5-acre) Outdoor Experiential Learning Site. The Outback is a student-run organic farm, community garden area, and restored wetland. There are trails that wind through the various sections of the site, including cultivated gardens and natural forested areas. As part of the Pacific Northwest maritime ecology, native botanical species in the Outback include Western red cedar (*Thuja plicata* Donn ex D. Don), Douglas fir (*Pseudotseuga menziesi* [Mirb.] Franco), red osier dogwood (*Cornus sericea* L.), and Oregon grape (*Berberis aquifolium* Pursh). I offer this workshop in the late spring (usually early May) when the local plants have emerged and many are in their flowering phase, which aids the identification process. I make use of the Outback's outdoor covered classroom area with bench seating for the students. This area is helpful for gathering students together for the medicine-making demonstration portion of the lesson, and is a welcome refuge from the ever-present possibility of rain.

The Learning Activity

Objectives

The *Local Medicinal Plant Identification and Herbal Medicine-Making* workshop is designed with three basic objectives: 1) to familiarize students with local flora; 2) to introduce the field of ethnobotany generally and medical ethnobotany specifically; and 3) to transform students' theoretical learning into applied experience in the short window of one class period (approximately 2 hours). The workshop happens after the students have been exposed to readings, lecture, and discussion (see Notes for recommendations) about the role and importance of medicinal plants in world cultures throughout history. This first portion of their (usually week-long) focus on medicinal plants is historical and theoretical. The hands-on, outdoor workshop aims to complement the earlier classroom learning with a practical and applied style of learning about medicinal plants. It shows students but one example of how their learning in the classroom unfolds in the "real" world and illustrates the relevance of the concepts and methods to their personal lives. The outdoor, hands-on workshop makes use of a nearby, outdoor experiential learning site where many native and naturalized plants are abundant, lending itself to the familiarization of students with local flora, which is a complementary aim of the Ethnobotany course. Furthermore, the workshop showcases concepts and methods of the practice of herbalism and follows up through suggestions of resources for optional further study.

Activity Narrative

The lesson on *Local Medicinal Plant Identification and Herbal Medicine-Making* begins by meeting in the entrance to the Outback/Outdoor Experiential Learning Site, a sort of community park, near Fairhaven College on Western Washington University's campus (the steps of the exercise are summarized in Table 4.1). At the forefront to the Outback is an Herb Garden that students have created and maintained over several decades. I meet the group in an open space where we gather in a circle for introductions. I welcome the students to the garden and the workshop. I introduce myself and offer my qualifications for leading the workshop—I am a Fairhaven alumnus, my interdisciplinary concentration from Fairhaven College is called "Herbal Medicine in Cultural and Ecological Contexts," and I grow an herb farm for my herbal products manufacturing business in a neighboring county. I next use one of several options for having the students introduce themselves:

1. Give your name and briefly explain your experience with herbs, if any.
2. Give your name and tell us your favorite plant (a tree, food, flower, etc.).
3. Give your name and the name of a plant that shares the same first letter as your name (e.g., Sally Sunflower, Amanda Artichoke).

Next, I bring the students' focus to the place surrounding us. The course has already introduced the Outback through other activities, but not all students are familiar with the Herb Garden. I explain the plan for the workshop: a plant walk followed by a medicine-making demonstration. I explain that we will begin by walking along trails through the Herb Garden and then meander farther into the small forest patch and through some other community garden areas. As we go, we will identify plants, discuss their medicinal uses, and harvest a

Table 4.1. Outline of steps of the activity.

Activity	Description
Introductions	• Presenter introduction • Student introductions • Introduction to the site/landscape • Plan
Guided plant tour	• Local botanical identification • Plant descriptions • Plant harvesting
Demonstration	• Plant preparation • Extract-making • Discussion of medicinal properties, and methods of manufacture and use • Share examples
Final discussion	• Review • Resources for further study • Questions

few plant parts. We will then gather in the outdoor classroom space for the herbal medicine-making demonstration, review, and discussion.

We begin the guided plant tour in the Herb Garden, which is a meadow-like area with approximately 93 m^2 (1000 square feet; ≈11-m [36-inch]-diameter circle) of raised beds and low-lying landscaped beds connected with wood-chip-covered paths. The Herb Garden has a high concentration of medicinal plants, which makes it an ideal area to identify plants in a group setting. I walk the students along the garden paths, pointing out the medicinal plants, one by one (e.g., Hops [*Humulus lupulus* L.], Wood Betony [*Stachys officinalis* (L.) Trevis. Ex Briq.], Valerian [*Valeriana officinalis* L.], Black Cohosh [*Actaea racemosa* L.], Elecampane [*Inula helenium* L.], Raspberry [*Rubus idaeus* L.]) After identifying the plant, I describe its principal medicinal actions (i.e., sedative, expectorant) and elaborate on what that means for its common uses within the Western herbal tradition. I mention the center of origin of the plant and any associated historic lore, and personal anecdotes about its uses.

We continue through a small forest patch and other community garden areas with this process of identification and description. In the forested area, we discuss native plants with medicinal uses (e.g., Oregon grape [*Berberis aquifolium*], nettle [*Urtica dioica* L.]). In the meadow-like areas of the community gardens we discuss naturalized plants with medicinal uses (e.g., Hawthorn [*Craetageus monogyna* Jacq.], Dandelion [*Taraxacum officinale* F.H. Wigg], Burdock [*Arctium lappa* L.]). I incorporate examples of herbal preparations, such as tinctures or salves, to illustrate the plants' uses or expand upon a related point. For example, in discussing Oregon grape roots' bright yellow chemical constituent berberine, I scrape a small portion of the plant's stem to show the yellow color, and then share a tincture (an alcohol extract) of Oregon grape root to show that the alcohol effectively extracts berberine (among other chemical constituents), including its characteristic color. If students opt to taste a drop of tincture—which I encourage—they understand, in their bodies, the description of the root as useful in promoting salivary secretions and aiding digestion because of Oregon grape root's bitter taste.

As we walk through the Outback, we harvest a small amount of certain plants along the way so that we have local plant materials to use in crafting herbal medicines of our own during the demonstration. I explain, and we discuss, considerations for safely and ethically harvesting wild plants, including the need to correctly identify plants, the importance of avoiding environmental pollutants, obtaining permission to harvest, and observing limits on harvest amounts. During the walk, we harvest from plants that are abundant and only the plant parts that regrow easily. We also harvest with a purpose in mind (the herbal medicine making demonstration). For our medicine, one or two students harvest lemon balm (*Melissa officinalis* L.) stalks and leaves, and leaves of dandelion, plantain (*Plantago major* L.), and cleavers (*Galium aparine* L.).

After touring the entire Outback, we arrive at the outdoor classroom. The students sit on benches arranged in a semi-circle and we put all of our harvest bags in the center. I explain the medicinal preparations that we will be making: an herb-infused honey, and an herb-infused vinegar. Students volunteer to demonstrate preparing the herbs, which I talk them through.

Instructions for the herb preparation are the following:

- Pick through the herbs, one by one, to remove any damaged parts.
- Chop up herbs to approximately 6.5 cm^2 (1 square inch). Be careful of stinging "hairs" on nettles, or other plants' spines or potential dangers. Wear gloves when necessary.
- Place herbs in jars. Pack them down gently so that air space surrounding the herbs is minimal, but not so tight that there is no space for the solvent to flow around the herbs. Leave about 5 cm (2 inches) of space from the top of the jar.
- In the jar with herbs suitable for an infused honey, cover herbs with honey and fill the jar to about 2.5 cm (1 inch) below the top of the jar (an inch above the herbs). Poke down the herbs with a knife, spoon, or chopstick, to make sure the honey completely mixes with the herbs and air bubbles are released.
- In the jar with herbs suitable for a vinegar extract, repeat the step above using apple cider vinegar instead of honey.
- Securely fasten lids to the jars.
- Give a gently shake to the jar to combine the herbs and solvent thoroughly.
- Label each jar with the date and contents.

As we go, I review the medicinal properties of each herb and we discuss potential uses of the herbal extract. We make a lemon-balm-infused honey to use, once complete, as a sweetener in tea or taken by the spoonful as a relaxant. We make a "spring greens" vinegar with nettle, dandelion, plantain, and cleavers, that is a mineral-rich, gently cleansing extract to use as an addition to drinking water or as an ingredient in salad dressing. I explain the process of medicine making and ask for volunteers to follow up on the straining process of our extracts once they have steeped sufficiently (about a month-long process).

Next, we discuss other medicinal preparations of herbs, such as teas, poultices, and salves. By now, students have witnessed the making of an herbal oxymel (honey extract) and vinegar extract, and have seen an example of an herbal tincture (the alcohol extract of Oregon grape root). We turn to descriptions of teas (infusion and decoctions), poultices, infused oils, salves, and essential oils. I bring examples of each type of preparation so that they can see, smell, and taste (when appropriate) or try the various ways of using herbs. For example, I explain:

- A tea may be an infusion or decoction. An infusion is made by steeping flowers, leaves, and/or stems of a plant in freshly boiled water for 5–20 minutes. A decoction is made by simmering berries, roots, and/or bark of a plant for 10–20 minutes. Each is then strained and ready to drink (or apply topically). Sometimes I bring a sample of lemon balm tea for students to try.

- A salve is an oil-based medicinal product for topical use that is made by steeping an herb in a carrier oil (olive, coconut, almond, etc.) and then straining it out. The oil extracts the medicinal constituents of the herb. Then the oil is heated, a small amount of beeswax is melted and mixed in, and then the liquid is poured into a container. When it cools it becomes solid and is thus easy to apply to the skin. I bring a sample of arnica (*Arnica cordifolia* Hook.) salve for sore muscles that students may try.
- An essential oil is a concentrated, volatile, aromatic oil found within plants. It is usually extracted by steam-distillation. Essential oils usually need to be diluted in a carrier oil (olive, coconut, almond, etc.) for safe use. I bring a small bottle of lavender (*Lavandula angustifolia* Mill.) essential oil that students may smell.

I provide a handout that defines and describes each herbal preparation so that students do not need to take notes.

Lastly, we review the herbs we harvested and the methods of preparation for creating honey and vinegar extracts (and instructions for finishing the process, since it cannot be completed in one day). I show a few books to encourage further reading and study, and suggest local and online resources for sourcing and learning about herbs and herbalism. I then open up the space for questions and comments, of which there are usually many.

Resources Needed

The first and foremost resources are the people involved: a course instructor or guest speaker with knowledge and experience with the topics, and an audience. Equally important is access and permission to use a nearby or easy-to-reach outdoor space with medicinal plants. Permission to harvest plants on the site is ideal, although without it, fresh or dried plant material may be brought from elsewhere for use in the demonstration. This is an equally viable option in situations where the health of the ecosystem is unknown, questionable, or downright subpar.

Using the outdoor space wisely is also necessary. For example, in areas with extreme heat or sun exposure, making use of shade could be indispensable to the wellbeing of the students and overall success of the workshop. In this particular case, having a sheltered classroom space is helpful in case of rain. Having an area to gather and sit as a group is also necessary for the demonstration portion of the workshop. Disability access is another issue to consider, as garden trails are often uneven and narrow, or may require climbing steps or stairs.

Other resources are needed to carry out the activities in the workshop, and are particular to the various participants. Students should wear appropriate clothing or gear for being outdoors in the local climate, and should bring along drinking water. The instructor or guest speaker should bring materials needed for plant harvesting and processing:

- Scissors or garden shears
- Plant collection bags
- Two 1-L (or 1-quart)-sized jars
- 1 L (or 1 quart) honey
- 1 L (or 1 quart) apple cider vinegar
- 2 or 3 sharp kitchen knives
- 2 or 3 large cutting boards
- pen and label paper to identify the finished products made in the demonstration

Optional resources include brewed tea (with cups) for students to try, various herbal products to use as examples (tinctures, salves, essential oils), and a handout listing and defining herbal medicine preparations. Books or plant guides may also be used for reference and as examples of resources for further study. Lastly, one or more volunteer(s) is required to hold onto and eventually strain the extracts that are made during the demonstration. After that process, they will bring the extracts back to the class to share the finished product with classmates, who may choose to bring in their own small bottles or jars to take some of the extracts home.

Outcomes

Student Accomplishments

This workshop accomplishes four primary outcomes. First, students achieve a familiarization with, and usually an enhanced appreciation for, useful medicinal plants and the natural ecology of their local environment. They get practice identifying local plants and make connections about plant communities and local ecology (e.g., nettles can be found growing underneath Douglas fir trees near a creek).

Second, students learn about ethnobiology and especially about medicinal ethnobotany. In addition to the readings, with each new plant that they encounter on the walk, they hear the stories of local peoples—Indigenous and settlers—and their interactions with the plants. In the cases of introduced plants, they will also hear of traditional plant-specific uses elsewhere. This will indicate the long and varied history and continuing appreciation and behaviors surrounding the plants, as well as reinforce the notion that ethnobiology is "inherently cross-cultural" (Lepofsky and Feeney 2013:35). Furthermore, ethnobiology's interdisciplinary nature makes it best studied and experienced in multiple contexts: in the classroom and outdoors (Lepofsky and Feeney 2013). This workshop takes students across the borders of anthropology, biology, history, and medicine, and into the natural classroom outdoors.

Third, students link theoretical and abstract classroom material (readings, lectures, discussions) with a relevant application of their learning in a way that engages all five senses. This experiential learning solidifies and expands "book learning" or "mediated knowledge" by illustrating the real-world applications of the information through a "direct encounter" (Rozzi et al. 2006:43). The final products created through the demonstration (the honey and vinegar extracts) serve as reminders of the learning process and of the information that students were exposed to within the workshop. Students learn basic information, in action, about the manufacture of herbal preparations.

Fourth, by engaging with each other outside of the normal classroom, students create deeper bonds with one another and a new sense of connection within their learning community. In addition to the tangible and pragmatic benefits of opening up new channels of collaboration, feedback, and support, connection aids their sense of belonging in that particular class, a demonstrable pedagogical advantage (Walton et al. 2011). The sharing in experiential exercises leads students to identify psychologically as a member of that course, which positively impacts student motivation.

Another outcome of this workshop is that, because of my personal experience as a Fairhaven student and alumnus, students see an example of how studies at Fairhaven can be useful in a related career. While this does not directly affect the learning students achieve in the workshop, it may enhance the students' awareness of their own educational path and possibilities. Of course, faculty who are not fellow alumni may nevertheless relate with their students in other aspirational ways in the course of the outdoor workshop.

Reflection

In examining the effectiveness of this workshop as a learning opportunity, two areas deserve discussion: the practical and the pedagogical. I share what worked and what did not, from my perspective and from what the course instructor and students have shared with me. Pedagogically, several things worked about this workshop. Splitter and Sharp (1995; see also Garrison et al. 2010) speak to the importance of creating a "community of inquiry" within an educational experience, which I feel to be relevant to understanding at least part of the success of this workshop. In this "community of inquiry" framework, three forms of "presence" interact to create a dynamic whole: social, instructor, and cognitive (Garrison et al. 2010). Social presence in this workshop begins with student introductions and encouraging active participation throughout the workshop, leading to the creation of a sense of community (students' previous shared experience throughout the larger Ethnobotany course notwithstanding). The "teaching presence" that I brought to the workshop included my interest in creating connections with students while still maintaining an authoritative role. I did this by engaging students personally and with each other in structured conversation and discussion, and by sharing information and experiences that illustrated my topical expertise. Presenting relevant, engaging information that complements previous learning in the course ushers student cognitive presence, both in terms of content (applications of plants) and presentation (hands-on, outdoor, experiential). The outdoor workshop intertwines these three forms of presence into a learning experience that is supportive, dynamic, and cohesive.

Primary school students generally respond well to outdoor and experiential learning experiences (Ballantyne and Packer 2002; Martin et al. 1981), especially when teachers nest these experiences within course curriculum (Dillon et al. 2006). It follows that undergraduate students also respond well to such experiences, evidenced in this case by the positive feedback I consistently receive after conducting the workshop. The applied nature of the workshop material complements the related "book learning" in the Ethnobotany course. It provides an opportunity for students to see knowledge in practice that has direct applications to their own lives and wellbeing. The workshop engages different styles of learning—and perhaps different "intelligences," including a naturalistic intelligence that makes sense of the natural environment (Checkley 1997; Gardner and Hatch 1989). Student engagement and enjoyment is high, as the workshop invokes multiple and active learning styles. The activity increases contact between and among instructor(s) and students compared to regular classroom settings, contributing to several of Chickering and Gamson's (1987) principles for good practice in undergraduate education. In this workshop, being outdoors engages all senses and seems to energize the students to pay extra attention to what is happening, even given the inherent distractions of the outdoor classroom (i.e., noise, weather, animal and human traffic).

The hands-on nature of this type of learning is important, not just for student satisfaction, but also because it instills an environmental ethic. How we talk about and understand nature has implications for what we see our role to be in relationship with it (Rozzi 1999). An educational program that uses first-hand experience, or "direct encounter" of nature "leads to a conceptual awareness of biological and cultural diversity, but also to an emotional feeling and sense of involvement" that motivates future actions, often in support of biodiversity conservation (Rozzi et al. 2006:43). Even through "small and incremental lessons," people may develop an understanding of nature that is imbued with an ethic of care for it (Turner and Berkes 2006:504). In this particular lesson, students learn to embody a conservation ethic by understanding—and, at the same time using—the natural resources around them.

Additionally, students' openness to and acceptance of the topic of medicinal plants is instrumental to the success of this workshop. In part, these students likely already have some level of familiarity and openness to natural health topics from living in the Pacific Northwest context, where environmentalism and alternative health practices are widespread. Student openness and acceptance is likely in this course because they already decided to enroll in Fairhaven College in general, and the Ethnobotany course in particular, and because of their learning experiences granted by those choices (although, openness to outdoor learning may be high in any course covering ethnobiology). Students have expressed satisfaction with being able to explore concepts and methods of herbalism within their local environment. Many indicated that they knew or believed that herbs have health effects, but that they had never had an opportunity to learn more specifics of herbs' actions or use.

Pedagogically speaking, something that did not work was attempting to cover too much information in the workshop. Throughout my years leading this workshop, I attempted, on at least two occasions, to include more than a handful of plant identifications and descriptions, highly involved theory about the manufacture of herbal preparations, or cross-cultural health system information and comparisons. Both the limited time frame (2 hours) and student response (decreased attention) informed me that this workshop is not the time or place for these extrapolations. Sticking with the basics (describing 10–12 plants, harvesting 4–5 of them, describing 5–6 methods of herbal extract preparation explained in Activity Narrative section) was quite sufficient, especially because this short workshop is solely an *introduction* to local medicinal plants and concepts and methods of herbalism. It is tempting to try to include as much subject matter as possible, but less is usually more in this regard.

Practically speaking, several factors worked well for this learning activity. Foremost was having an appropriate nearby outdoor site in which to meet. The ease with which students reached the site (a five-minute walk from their normal classroom) meant that the workshop was accessible and convenient, which was important to its feasibility in terms of time and minimal transportation resources required (for more on the importance of ease in creating learning opportunities on "green school grounds", see Dyment 2005:39). The site is appropriate in the sense that it has a range of different plants to identify and discuss, and an outdoor classroom which allows a level of comfort and cohesion for group meetings in an outdoor space. Elsewhere, campus gardens or greenhouses may serve as alternatives.

The group size of 20–24 people (20–22 students, the Ethnobotany course instructor, and myself) was manageable but not ideal. Finding space for the participants to stand and gather around garden beds, as well as the ability for our voices to be heard over the elements of the outdoor classroom, were at times challenging. A smaller group of 12–15 students would have reconciled those issues and focused the energy and attention of the students more. On the other hand, the large group supports a high degree of energy and enthusiasm, which is motivating for the learning activity.

Other practical considerations that worked well involve the organization and logistics of the workshop. The workshop was held during regular class time, so that students need not plan for it outside of their normal schedule and responsibilities. The class was always scheduled in the mid-to-late afternoon, which meant that students usually had few time-commitments immediately following the workshop. This slot allowed optional time for extra questions, comments, and discussion following the workshop. The course instructor supplied some materials (jars, honey, vinegar) and I brought others (cutting boards, knives, harvest bags, label paper, pens, examples of herbal products) which seemed practical and cost-effective. Western provided me an honorarium (stipend) for the workshop, which helped defray my costs of time and transportation. My familiarity with the course material (having been a student) and my rapport with the Ethnobotany course instructor (a former academic advisor) both helped in planning, logistics, and content preparation. Lastly, offering resources for optional further study (by showing books and recommending local shops and community teachers) was helpful to students' continued learning and their motivation for exploring the topics more.

Conclusions and Future Prospects

Situating the Lesson within the Course and Institution

The *Local Medicinal Plant Identification and Herbal Medicine-Making* lesson I have outlined in this chapter is both distinct from, and related to, other learning opportunities within this institution. As a hands-on, outdoor, experiential workshop, it is not a typical classroom lesson, in terms of location, activities, or style of presentation. Students engage with each other and with their instructor(s) in a new setting and with new topics. These topics, however, relate to aspects of the lessons within the Ethnobotany course, in which there is obviously a focus on the relationships of people and plants. Furthermore, the Ethnobotany course allows students other opportunities for outdoor engagement: local plant identification, collection for a plant press, and a gardening project. As such, this workshop fits in nicely with the course's other activities. Three prominent messages I have received as feedback from students are that 1) they love going outside for class, 2) they appreciate having a guest speaker as a change to the normal pace of their course, and 3) they appreciate introduction to knowledge they can put to use in their personal lives. The workshop also complements other Fairhaven College and Western Washington University courses, and general aims of the institutions, which encourage experiential and applied study at all levels.

Applying this Workshop in Other Courses and Institutions

There is increasing awareness about the benefits of engaging people in outdoor learning (see Bogner 1998; Kellert and Wilson 1995; Nabhan and Trimble 1995; Wattchow and Brown 2011), and relatedly, a rise of ethnobiology and human ecology topics or courses in institutional curriculums. The *Local Medicinal Plant Identification and Herbal Medicine-Making* workshop I have outlined in this chapter can contribute further to the rise in outdoor, experiential, ethnobiological education and to the development of an environmental ethic, by being adopted or adapted to other institutions.

To apply this workshop in other institutions, I suggest several considerations. First and foremost is to situate the workshop within a larger course or program that provides context for the topics or related knowledge. Second, assess the culture or social climate of the student body, the institution, and the local area in which this workshop might be offered. Is there interest in the topic and format? Third, what are the resources available to organize this workshop? Does the institution have a system already in place to provide support for materials and/or guest speakers? Is there a suitable outdoor space in which to hold the workshop? Is there a qualified speaker interested in collaborating? If not, how could one be recruited? Lastly, I would add that hosting the workshop regularly over a span of time helps provide experience and feedback for adjusting and refining the lesson aims and activities, and contributes to the ease with which it is organized.

Undoubtedly, this lesson will look different when conducted in different situations. The particularities of the local environment, the course in which it is embedded, and the people involved, will influence how the lesson unfolds. However, the basic structure of the lesson that I outlined in this chapter should function as a roadmap in adopting or adapting the lesson to other contexts. The important thing is to get people outdoors and into the garden. That's where the magic happens.

Notes

Suggestions for related lesson materials:

Brill, S., and E. Dean. 1994. *Identifying and Harvesting Edible and Medicinal Plants in Wild (and Not So Wild) Places.* Hearst Books, New York (especially Chapter 1: An Introduction to Foraging for Wild Plants).

Green, J. 2011. *The Herbal Medicine-Maker's Handbook: A Home Manual.* Crossing Press, Berkeley, CA (especially Chapter 1: A Field Trip, and Appendix B: Welcome to the Garden).

Heinrich, M., and S. Gibbons. 2001. Ethnopharmacology in Drug Discovery: An Analysis of Its Role and Potential Contribution. *Journal of Pharmacy and Pharmacology* 53(4):425–432.

Plotkin, M.J., dir. 2001. The Shaman's Apprentice. [Film]. Miranda Productions, Boulder, CO.

Swerdlow, J.L., and L. Johnson. 2000. *Nature's Medicine: Plants that Heal: A Chronicle of Mankind's Search for Healing Plants Through the Ages.* National Geographic Society, Washington, D.C.

Tyler, V.E. 1996. Natural Products and Medicine: An Overview. In *Medicinal Resources of the Tropical Forest: Biodiversity and Its Importance to Human Health,* edited by M.J. Balick, E. Elisabetsky, and S.A. Laird, pp. 3–10. Columbia University Press, New York.

Acknowledgments

I wish to thank Dr. John Tuxill and his Ethnobotany students for the many opportunities to share my knowledge and passion for plants and people via this workshop. To John, I owe my deepest appreciation for guiding and encouraging my ethnobotanical studies. Thanks to Dr. Gary Bornzin for suggesting that I teach without any formal pedagogical preparation so as to keep my originality and enthusiasm, and to Laurel Vukovic for inspiring me and for her steadfast mentorship of my herbal studies. I also thank Dr. Marsha Quinlan profusely for her thoughtful and constructive input on this chapter.

References Cited

Ballantyne, R., and J. Packer. 2002. Nature-Based Excursions: School Students' Perceptions of Learning in Natural Environments. *International Research in Geographical and Environmental Education* 11(3):218–236.

Bogner, F.X. 1998. The Influence of Short-term Outdoor Ecology Education on Long-term Variables of Environmental Perspective. *The Journal of Environmental Education* 29(4):17–29.

Checkley, K. 1997. The First Seven . . . and the Eighth: A Conversation with Howard Gardner. *Educational Leadership* 55:8–13.

Chickering, A.W., and Z.F. Gamson. 1987. Seven Principles for Good Practice in Undergraduate Education. *AAHE Bulletin* 3:7.

Dillon, J., M. Rickinson, K. Teamey, M. Morris, M.Y. Choi, D. Sanders, and P. Benefield. 2006. The Value of Outdoor Learning: Evidence from Research in the UK and Elsewhere. *School Science Review* 87(320):107.

Dyment, J.E. 2005. Green School Grounds as Sites for Outdoor Learning: Barriers and Opportunities. *International Research in Geographical & Environmental Education* 14(1):28–45.

Gardner, H., and T. Hatch. 1989. Educational Implications of the Theory of Multiple Intelligences. *Educational Researcher* 18(8):4–10.

Garrison, D., T. Anderson, and W. Archer. 2010. The First Decade of the Community of Inquiry Framework: A Retrospective. *The Internet and Higher Education* 13(1–2):5–9.

Green, J. 2011. *The Herbal Medicine-Maker's Handbook: A Home Manual.* Crossing Press, Berkeley, CA.

Kellert, S.R., and E.O. Wilson, eds. 1995. *The Biophilia Hypothesis.* Island Press, Washington, D.C.

Lepofsky, D., and K. Feeney. 2013. Ten Principles of Ethnobiology: An Interview with Amadeo Rea. In *Explorations in Ethnobiology: The Legacy of Amadeo Rea*, edited by M. Quinlan and D. Lepofsky, pp. 34–46. Society of Ethnobiology, Denton, TX.

Martin, W.W., J.H. Falk, and J.D. Balling. 1981. Environmental Effects on Learning: The Outdoor Field Trip. *Science Education* 65(3):301–309.

Nabhan, G.P., and S. Trimble. 1995. *The Geography of Childhood: Why Children Need Wild Places*. Beacon Press, Boston, MA.

Rozzi, R. 1999. The Reciprocal Links Between Evolutionary-Ecological Sciences and Environmental Ethics. *Bioscience* 49(11):911–921.

Rozzi, R., F. Massardo, C.B. Anderson, K. Heidinger, and J.A. Silander Jr. 2006. Ten Principles for Biocultural Conservation at the Southern Tip of the Americas: The Approach of the Omora Ethnobotanical Park. *Ecology and Society* 11(1):43.

Splitter, L.J., and A.M. Sharp. 1995. *Teaching for Better Thinking: The Classroom Community of Inquiry*. Australian Council for Educational Research, Melbourne, Australia.

Turner, N.J., and F. Berkes. 2006. Coming to Understanding: Developing Conservation Through Incremental Learning in The Pacific Northwest. *Human Ecology* 34(4):495–513.

Walton, G.M., G.L. Cohen, D. Cwir, and S.J. Spencer. 2011. Mere Belonging: The Power of Social Connections. *Journal of Personality and Social Psychology* 102(3):513–533.

Wattchow, B., and M. Brown. 2011. *A Pedagogy of Place: Outdoor Education for a Changing World*. Monash University Publishing, Melbourne, Australia.

Chapter 5.

The Mandala Exercise for Increasing Ecological Understanding

Steve Wolverton[1]

Chapter Summary

It is difficult in the typical college classroom to encourage opportunities that increase ecological understanding, or values resulting from direct encounters with outdoor environments. I adapted the approach developed by David Haskell in his popular book, *The Forest Unseen: A Year's Watch in Nature* (2013), for ethnobiology. In his book, Haskell describes his recurring visits to a square meter of an old-growth Appalachian forest in eastern Tennessee, which he termed his mandala—a word he used to metaphorically describe continual processes of ecological change in a place. Mandala exercises were set up for students such that they would establish routine visits to small outdoor places, to record observations in terms of biota, place, and people (including themselves) and to note changes over time. Each week, students might have chosen a biotic theme, research it as a thread, and then compose a short blogpost about their experiences. Alternatively, they may have simply recorded human-biota interactions in their mandalas. I found that student engagement and writing were exceptional, and I describe the outcomes of the exercise in this chapter.

Introduction

The mandala exercise comes from *The Forest Unseen: A Year's Watch in Nature* by biologist, David Haskell (2013). He had his students select a square meter of space in the outdoors

1 Department of Geography and the Environment, University of North Texas

and describe biological processes taking place there. In his book, Haskell adopts a similar approach by selecting a square meter of old growth forest in the southern Appalachians. He visited his "mandala" multiple times per month, chose a biological thread of inquiry to observe and investigate, and wrote engaging essays about what he experienced and learned. I read this book in the fall of 2013 while making plans to teach ethnobiology the following year, and I decided to use the exercise to address a few fundamental challenges I was facing in the classroom.

First, I find upper division undergraduate seminars to be frustrating to teach because many students do not read to prepare for class discussion. As a result, conversations in class do not go as deep into the material as I would like. My education was interdisciplinary and covered a lot of ground from language and culture to environmental chemistry, but my home field is zooarchaeology, which is the study of animal remains from archaeological sites to understand past human-environment interactions. I love zooarchaeology, but I had stopped teaching it as a stand-alone course because the depth of discussion and reading by students was so limited that I was chronically disappointed in how the course went. I chose to teach ethnobiology because it covers human-environment interactions in multiple contexts, which opens up a much broader array of readings, contemporary issues, and conversations with students. At my institution, there is a lot of freedom to design seminars around special topics (e.g., zooarchaeology, ethnobiology, or another topic), which may include courses that are reading-heavy and discussion based, blends of lecture and discussion, and even field trips. The seminars that I teach tend to be reading and discussion oriented. With ethnobiology, I did not want to go down the same path I had taken with zooarchaeology and other upper division seminars; I wanted to have part of the course be field driven and engaging for students. The mandala assignment was a good exercise for meeting this goal, and it was my hope that if I designed the course to link readings and discussion questions to reflexive themes to be considered during mandala observations that a higher proportion of students would read deeply and comprehend more. This turned out to be the case.

Second, I was also dissatisfied with the level of essay writing from previous upper division undergraduate seminars, including zooarchaeology, other archaeology courses, and seminars that centered on aspects of human ecology. The norm had become that one or two excellent students would thrive in those seminars, and some of those students later became my graduate students. However, at least half of the students were poorly prepared to write essays on the course material. Essay composition, in many ways, is simply another form of conversation between the students and I. In some classes, a large number of students were copying narratives directly from source materials, unknowingly or knowingly committing plagiarism. The university solution to this type of problem seemed to rely on increased investment in faculty policing of students. The situation disappointed me to the point that I began only assigning in-class writing assignments, discarding the old mainstay of term papers altogether. As with in-class discussion, the experience was dissatisfying; on one hand, I was often upset with students, on the other, I felt like I was not engaging students well. I felt like my teaching style was increasingly isolated into a bland corner.

These first two challenges relate more broadly to the downside of "passive learning" that takes place in most classrooms, particularly via the lecture format (Abeysekera and Dawson 2015; Andrews et al. 2011). The mandala exercise is an example of "active learning," which takes place when "an instructor stops lecturing and students work on a question or task designed to help them understand a concept" (Andrews et al. 2011:394). The mandala exercise thus "flips the classroom" (Berrett 2012), which entails reversing expectations of the traditional lecture, such that active learning is emphasized (Abeysekera and Dawson 2015). The flipped classroom model is widely considered to be an innovative pedagogical approach with positive impacts on learning outcomes in a wide variety of disciplines (e.g., Gilboy et al. 2015; McLaughlin et al. 2014; Roach 2014), though the impacts of such approaches are only recently being evaluated across higher education (Abeysekera and Dawson 2015; Andrews et al. 2011; Betihavas et al. 2016). With the mandala exercise, I decided to flip the classroom to determine if student engagement in reading from, discussion of, and writing about the scholarly literature in human ecology improved, but I was also interested in seeing if such an approach would lead to deeper discussion of environmental values, which is the third challenge I faced.

During my graduate work I had taken environmental ethics and was inspired by the well-honed argument of Ricardo Rozzi (1999) and others, that direct field experience in the outdoors has the greatest positive impact on environmental values. That is, if environmental scientists and social scientists want to influence values in society, the best way to do so is through experiential (or active) learning. This is not news to the ethnobiologist (Hunn 2014). Turner and Berkes (2006), for example, reflect upon cross-cultural differences in conservation ethics, illustrating that members of many Indigenous societies who spend time in nature show deep ecological understanding. When I read the ethnobiological literature, such as Gary Nabhan's or Gene Anderson's many books and articles (e.g., Anderson 2010, 2014; Nabhan 2004, 2016), ecological understanding is clear and inspiring, but that might not be so for students. I suspected that Rozzi, Turner, and others were correct—for many people it is direct encounters with the outdoors that leads to deeper ecological understanding. My thought was, "if I design a direct encounter experience, such as the mandala exercise through flipping the classroom, I bet students will get more out of the readings." I also presumed that the ethical implications of the experiences, readings, and discussions would more easily surface. It turned out that I was correct about how the mandala exercise would impact students. Weaving the exercise through the class changed everything.

In this chapter, I describe the mandala exercise and how I implemented it in my ethnobiology classes, and I summarize how discussion and essay writing improved. I also share excerpts from student essays and summarize feedback by students. I do not provide a detailed assessment of student outcomes because I did not design an assessment tool for the class;

instead I portray what I think worked well from the instructor point of view, acknowledging that a full assessment would have to be done to truly capture student perspectives on the course. That said, all anecdotal feedback from students that I have received supports that the experience was very meaningful for them; I present some of that feedback. I end the chapter by discussing the benefits and limitations of the exercise.

The Mandala Exercise

The ethnobiology course I teach takes place during a 15-week semester, either September through early December or January through early May. Early in the semester, students were to locate an outdoor place, roughly one square meter, that they would visit once per week. Some semesters students established mandalas off campus; other semesters mandalas were on campus. For example, when we did this on campus, during the first class period we went outside the UNT Environmental Science Building, and students walked the surrounding area. The building sits within a biodiversity hotspot on campus, which is partially the result of a surrounding post oak (*Quercus stellata*) grove with many large trees, but is also the result of native landscaping near the building. Just outside the Environmental Science Building, students were able to encounter a number of native trees, grasses, and other plants that are common in north Texas. In addition, there is some animal wildlife, such as fox squirrels (*Sciurus niger*), numerous birds, in addition to bronze statues of fauna that are native to the area. A condition was that mandalas had to be within 100 m of the building; students were to free to select the location of their mandala, but once the place was determined, it was set for the semester. That is, I made students aware that they were committing to observing the same place each week for the semester. Students were required to bring a separate journal notebook for recording observations. Each class period, one half hour of class time would be spent making observations in the mandala; this amounted to about one hour out of three per week that students made mandala observations.

Ideally, students would have a background in biology in an ethnobiology class, but many students had backgrounds solely in the social sciences as they may not have yet taken their core science courses. As a result, it was important to make it clear to students that each week, they were to select "something happening in or around their mandala" that struck them as new and interesting, perhaps looking inward to the mandala but at times looking outward from it. Rather than move from one observation to another, students were to follow a single process (e.g., a type of interaction with one plant or animal species or another) as far as possible. I used the Mandala Exercise for my ethnobiology classes during two semesters, for which I taught it slightly differently. One semester students selected a mandala off campus; another semester, I had students select a space on campus near the Environmental Science Building. Other than this difference, the instructions for the assignment were as follows:

> Students will select a mandala to visit multiple times per semester. Each field session, students will visit their personal 1 x 1 m mandala and make journal observations concerning environmental and cultural change for one half hour. Students will not manipulate the plot, following Haskell, and will simply take notes about observations made during the session (examples of observations are described below). After the observation period, students will spend time inside and outside of class learning about the processes they observed using online and scholarly sources; research related to observations is generally open-ended, but please ask for ideas and guidance as needed. Students will select one observation to research and will compose an essay about what they learned that will be turned in by the end of the week. Over the course of the semester, students will choose some of these essays to be published on meangreenmandala.blogspot.com. Students will combine their observations with more details from research online during the final portion of class, which will be cited in the blogpost.

Each week, I assigned three to four readings for discussion and periodically gave short lectures on important concepts related to that week's topic. We read one of Haskell's essays from *The Forest Unseen* (2013) every week in combination with other works in ethnobiology, such as selections from Gary Nabhan's book, *Cross-pollinations: The Marriage of Science and Poetry* (2004), Sutton and Anderson's *Introduction to Cultural Ecology* (2013), and/or scholarly papers published in the *Journal of Ethnobiology*, *Ecology and Society*, *Human Ecology*, or related journals. I framed discussion questions such that scholarly readings were conceptually linked to readings from *The Forest Unseen* (Haskell 2013), and I directed some part of every discussion toward linking students' personal mandala observations to the readings.

Outcomes

Mandala essays have varied over the years; at one point, we did not establish mandalas on campus. That year, essays covered a high variety of contexts because students in the class came from a diverse set of backgrounds. One non-traditional student lived in a suburb at the edge of rural north Texas, and his mandala was near an old farm pond in between two neighborhoods. A couple of students lived with their parents and wrote about places in their yards; thus, they embedded memories in their narratives. One student established her mandala behind her place of work (in a shop in a strip mall); she simply found a small patch of vegetation and strived to look close and then closer over time. Dana Lepofsky, at Simon Fraser University, adopted a similar approach, and her students compiled equally compelling essays on places they visited over the course of many weeks.

Despite the success of the off-campus mandala exercise, I decided one semester to have students do their observations during class time near the UNT Environmental Science Building. I made this change for a few reasons; first, I wanted to take advantage of students' enthusiasm for the exercise while in class. Second, I wanted students to spend time doing research in class online immediately after making observations, so that I could help guide them. Third, I wanted to more directly link mandala observations for the day to discussions of readings. Bringing the mandalas on campus accomplished these three goals, which led to more thorough research on mandala observations, better comprehension of class readings, and deeper discussions of ethnobiology that more clearly linked students' personal experiences to what they were learning in the literature. However, the cost of moving on campus was that the diversity of and personal connections to the mandalas was somewhat diminished.

The student products of the mandala exercise included a notebook (or journal) of their field observations. They used those notes during research to craft weekly essays about their observations. Students chose the best of their essays as their final product, which was published as a blogpost. Below, I draw quotes from those posts to illustrate student outcomes from the Mandala Exercise.

Students described a shift in their ecological understanding or environmental awareness in their blog posts. Common themes included coming to view the world as entangled networks of abiotic and biotic interactions, including humans, other animals, plants, soils, atmospheric conditions, and other components of the environments. For example, Student A concluded:

> The interrelationships of animals, plants, and people are often overlooked and understated in our technology-and popular-culture-dominated world, but in reality humans are always acting within an ecosystem, whether they are aware of it or not, and it can only be advantageous to be aware of these interactions, especially within the context of conservation where knowledge can lead to the expansion and preservation of endangered biota and increased biodiversity.

Another common shift was an increase in awareness of surroundings; students often remarked how when they slowed down and paid attention during their observation periods, they became aware of details. For example, off-campus observations were imbued with a sense of place and memory for Student B during the semester that he took the class:

> Mandala, what an apt description of my little circle. I've sat here repeatedly almost every morning over the last two years and contemplated everything from marriage, to my children, to unemployment...

It was particularly rewarding as an instructor to encounter concepts from class readings integrated into the blog posts. Ecological understanding as defined by Turner and Berkes (2006:497) was a core theme that showed up in observations made by students. Reflecting upon feral hogs, for example, Student B concluded:

> Nancy Turner and Fikret Berkes (2006) use the term ecological understanding to refer to a set of attributes embodied within traditional ecological knowledge systems. The first of these attributes involves an incremental learning of environmental knowledge through detailed observation. In my observations . . . , I can only speculate due to being a complete novice, on the role each species (plant and animal) plays in maintaining or even modifying local natural resources. As an example, I knew that pigs were an invasive species, however through observation and prolonged deduction, I was able to build a picture, not only of the destruction caused by their feeding, but also the role they may play in spreading cockleburs, another invasive species. Pigs feeding in the same pastures as cattle, spread this poisonous plant, which might be ingested by domestic grazers. The pigs have only one other predator in this region besides humans, which is the coyote.

Another common theme that came through in student essays is that they took detailed notes during field sessions, echoing that slowing down and taking time to observe led to greater awareness and a sense of place. For example, Student C wrote:

> I returned to the flower bed in the afternoon to make some more observations. I looked closely at the rosemary bush to see if there were any changes since this morning. The temperature outside was the familiar warmth of a Texas afternoon. The sun periodically flashed its rays between the scattered clouds. Two bees were still hovering around the outside of the rosemary bush. I could not tell if these bees were the same ones that I saw this morning, but I did determine by color and size that they belonged to the same honeybee family. They continued to search for pollen, while avoiding each other as I took a closer peek into the bush. The small green lizard I found in the rosemary bush this morning was gone from sight. However, I caught a glimpse of a green tail between the rosemary branches, located near the center of the bush and away from danger. Another missing rosemary resident was the large green grasshopper that had been resting in the center of the bush. This large grasshopper was gone, but in its place I found a smaller and leaner yellow grasshopper. This grasshopper had yellow hind legs with dark bands and a dark brown abdomen. The grasshopper perched

> on a rosemary branch that had access to the sunlight. Slowly, the grasshopper would angle its body to gain more access to the sun. Though searching for sunlight and heat, the grasshopper stayed observant of outside dangers, including birds and other enemies, myself included. Speaking of birds, I did not see any fly near the flower bed mandala.

Student C concluded her mandala observations and blogpost with:

> One of the many benefits of making my own observations and reading Haskell's book is that I made more detailed observations of how the insects and animals interact in the miniature ecosystem of my mandala. I think this idea can be applied to humans around the world. It is important to observe and interact with local natural environments. These relationships with the environment can reveal crucial information about ... the multiple species of biota, which could have beneficial effects and qualities for human life and culture.

These are but a few of the many types of observations that students made during their mandala field, research, and journaling sessions. I last taught ethnobiology in 2015, and unfortunately, I did not retain the qualitative comments from student evaluations for my ethnobiology courses. However, class size tended to be small at roughly 10 to 12 students. I recall worrying that students would feel that too much had been asked of them through the mandala exercise. To the contrary, students were generally thrilled with the experience. I reached out to a few of the students recently and asked them to reflect on the exercise. Two students responded (not Students A, B, or C above).

One student who entered an applied anthropology master's program the semester after he took ethnobiology at UNT responded:

> I actually spend a lot of time thinking about how meaningful those kinds of exercises were and how impactful they have been on my current scholarship. A lot of the lessons of ethnobiology/environmental philosophy I took for granted. It's kind of crazy how observing plants and animals is a radical act these days but it is true. Especially if you try to bring in any discussion on environmental ethics. Which a lot of people, professors included, imagine as a separate "environment/nature" from everyday people. They can't image every day people having the capacity for environmental ethics that can link spirituality and cutting edge science.

A second student who did the exercise during a semester during which mandalas were established off campus responded:

> I was actually thinking about that the other day. When I sit outside with my dog, I still find myself looking down at the grass, or watching the ants crawl along the cracks in my cement patio. I hear dragonflies and watch where they're heading; catch glimpses of birds and wonder what kind they are and where their nests are. Last fall, I had a pair of hawks who frequented the squirrels in my yard, and once spent three hours sitting at the window watching them hunt. I've had a mama bird make a nest and lay eggs in the support beams of my carport for the last couple of years, and check each day to see how the brood is coming along, until they all grow up and take off. It's been 4 years, and I still look for the colors in the fall, the insects in the grass, the different kinds of plants growing wherever I am. I still find myself more observant after doing that project.

In summary, the short term and long term impacts of the mandala exercise appear to have been profound for students. In general, during semesters that I assigned the mandala exercise students read deeply, enthusiastically participated in discussion, and exhibited better writing than in my other classes. I discuss how I think this happened below.

Discussion

I have written this chapter from the perspective of my role as the professor, thus the implications of student outcomes are also filtered through my perceptions. I outlined three reasons that explain why I adopted the mandala exercise in the introduction—1) dissatisfaction with student preparation, mainly concerning reading and comprehension; 2) concerns about essay writing, primarily related to quality but also to the high potential for copying/plagiarism; and 3) a desire to resonate with students concerning environmental values, ones that are inherent in the scholarly literature, but that students were not internalizing from class discussions. I return to those reasons below, but there is a core change that I saw in students that permeated all three reasons—because the students chose their mandala and wrote about their experiences (the final product of which was shared publicly as a blog), they simply had more ownership over all aspects of the class.

Students read the material and showed particular enthusiasm for Haskell's book. In addition, because I explained how other readings link to Haskell's writing and their mandala work, students took the other readings seriously. Some of those readings might be characterized as traditional "stuffy" scholarly pieces, but students had a clearer *a priori* reason to wade through them. There were unexpected benefits to this; for example, students often did not understand parts of one or another scholarly work, but because they gave reading a more serious attempt, they came to class with questions. Many times, other students had meaningful answers to those questions. Discussion became what it is supposed to be, a place to grow understanding. In previous courses, without the investment in reading students were coming to class either poorly prepared to make contributions, nervous that she/he/they might have "the wrong answer", or a combination of both. Then as well, a few individuals would dominate discussions (unless I shut them down), and often they were posturing with what they perceived to be impressive answers. Such displays disappeared the semester I used the mandala exercise because most students were reading most of the time. What developed were enriching, honest conversations—there is no greater treasure as a professor than to be involved in these types of conversations with students. Such discussions took place all semester.

Essay writing was substantially improved compared to any other class that I teach. There were several aspects to this improvement. First, I encouraged students to write in their own voice and to not worry about perfection. It was uncomfortable, but I felt it was necessary to compromise on academic precision; many of the students had no training in biology, for example, and did not know taxonomy. To be honest, there was a lot of taxonomy that I did not know; however, my role became to guide students toward sources that could improve their understanding while still encouraging to write in their own terms (even if they were imprecise or at times incorrect). This, combined with the fact that they were 1) writing about their own observations, and 2) charged with researching the processes they observed, seemed to give them both freedom to write but also a focus within which to explore. Another important factor was that we published essays on the course blog. I did not anticipate that students would think of their posts as something to share with friends and family members. Thus, they were not writing essays in order to finish assignments, they were writing to share what they learned; this internal shift in audience from professor to public was a key element of the course.

It is more difficult to gauge how the course engaged students at the level of environmental ethics and ecological understanding. I am certain that students learned a lot of new ethnobiology, ecology, and biology because of the course readings and discussion as well as their own explorations related to their mandala observations. It is more difficult to determine whether or not this influenced the environmental values that students hold. Students who gravitated to the course in the first place often held interests in ecological anthropology or human-environment geography. Nonetheless, I assume that whatever environmental values that students held related to biocultural conservation were reinforced by the course. In addition, there were a few students who had no experience related to human ecology prior to the course, and for them the mandala exercise, the readings, essay writing, and discussions appear to have been foundational. Because I did not design the course with assessment of short or long term outcomes in mind, my understanding is limited to the few observations I made about student responses above and my perceptions as summarized here.

Conclusion

When I reflect upon the mandala exercise several considerations come to mind. First, why did it take so long for me to figure out that active learning bolsters comprehension of readings, participation in discussion, and clarity of writing? Clearly, there is pedagogical research that supports such a change (e.g., Abeysekera and Dawson 2015; Roach 2014). I wonder if other professors are as slow to change as I was. My answer is that I had been comfortably situated within my teaching habits, which included lecturing and seminar-oriented discussion classes—a passive learning model. I still find those types of courses comfortable, and in some contexts lecturing, for instance, is important and necessary. Adopting the mandala exercise felt like a risk because I wondered if I was abandoning important course content and discussion/lecture time by emphasizing the experiential component. It is no coincidence that I adopted the exercise only after I had received tenure; in short, the perceived risk was diminished enough to give it a shot.

Adopting the mandala exercise led me to re-conceptualize teaching in general. The mandala exercise worked well in my ethnobiology course, and I have worked to integrate active learning opportunities into all of my classes, even the ones that are large lectures. In earth science, a large introductory general education science course, I revised all of my lectures to make each period more conversational. Further, I regularly communicated to students that they are part of an "imagined conversation" much like a researcher is in her/his/their scholarly writings (Booth et al. 2016:16). I decided, for example, to make the exams in that course open-notebook. I spend lecture time teaching note-taking, and I give bonuses to students who take a note-taking workshop through my university's learning center. One message I deliver is that note-taking is a form of communication; it is an active learning process, if a person chooses to engage in it (Bui and Myerson 2014). In contrast to simply sitting in lecture and listening to lecture, which is more like watching a video, note-taking requires listening to, filtering, organizing, summarizing, and communicating through writing, which is a much deeper cognitive process (Mueller and Oppenheimer 2014). In summary, the mandala exercise impacted all aspects of my teaching. Using the exercise led to beneficial impacts related to reading, writing, and discussion in my ethnobiology seminar, but it also led me to work active learning pedagogy into my other classes.

References Cited

Abeysekera, L., and P. Dawson. 2015. Motivation and Cognitive Load in the Flipped Classroom: Definition, Rationale and a Call for Research. *Higher Education Research & Development* 34:1–14.

Anderson, E.N. 2010. *The Pursuit of Ecotopia: Lessons from Indigenous and Traditional Societies for the Human Ecology of Our Modern World.* Praeger, Santa Barbara, CA.

Anderson, E.N. 2014. *Caring for Place: Ecology, Ideology, and Emotion in Traditional Landscape Management.* Left Coast Press, Walnut Creek, CA.

Andrews, T.M., M.J. Leonard, C.A. Colgrove, and S.T. Kalinowski. 2011. Active Learning Not Associated with Student Learning in a Random Sample of College Biology Courses. *CBE—Life Sciences Education* 10:394–405.

Berrett, D. 2012. How 'Flipping' the Classroom Can Improve the Traditional Lecture. *The Chronicle of Higher Education* 12:1–3.

Betihavas, V., H. Bridgman, R. Kornhaber, and M. Cross. 2016. The Evidence for 'Flipping Out': A Systematic Review of the Flipped Classroom in Nursing Education. *Nurse Education Today* 38:15–21.

Booth, W.C., G.G. Colomb, J.M. Williams, J. Bizup, and W.T. Fitzgerald. 2016. *The Craft of Research*, 4th edition. University of Chicago Press, Chicago, IL.

Bui, D.C., and J. Myerson. 2014. The Role of Working Memory Abilities in Lecture Note-taking. *Learning and Individual Differences* 33:12–22.

Gilboy, M.B., S. Heinerichs, and G. Pazzaglia. 2015. Enhancing Student Engagement Using the Flipped Classroom. *Journal of Nutrition Education and Behavior* 47:109–114.

Haskell, D.G. 2013. *The Forest Unseen: A Year's Watch in Nature.* Penguin, New York.

Hunn, E. 2014. To Know Them is to Love Them. *Ethnobiology Letters* 5:146–150.

McLaughlin, J.E., M.T. Roth, D.M. Glatt, N. Gharkholonarehe, C.A. Davidson, L.M. Griffin, D.A. Esserman, and R.J. Mumper. 2014. The Flipped Classroom: A Course Redesign to Foster Learning and Engagement in a Health Professions School. *Academic Medicine* 89:236–243.

Mueller, P.A., and D.M. Oppenheimer. 2014. The Pen is Mightier than the Keyboard: Advantages of Longhand over Laptop Note Taking. *Psychological Science* 25:1159–1168.

Nabhan, G.P. 2004. *Cross-Pollinations: The Marriage of Science and Poetry.* Milkweed Editions, Minneapolis, MN.

Nabhan, G.P., ed. 2016. *Ethnobiology for the Future: Linking Cultural and Ecological Diversity.* University of Arizona Press, Tucson.

Roach, T. 2014. Student Perceptions toward Flipped Learning: New Methods to Increase Interaction and Active Learning in Economics. *International Review of Economics Education* 17:74–84.

Rozzi, R. 1999. The Reciprocal Links between Evolutionary-Ecological Sciences and Environmental Ethics. *BioScience* 49:911–921.

Sutton, M.Q., and E.N. Anderson. 2013. *Introduction to Cultural Ecology*, 3rd edition. Altamira Press, Lanham, MD.

Turner, N.J., and F. Berkes. 2006. Coming to Understanding: Developing Conservation through Incremental Learning in the Pacific Northwest. *Human Ecology* 34:495–513.

Chapter 6.

Using the Management of Urban Trees to Teach Conservation Biogeography

Lisa Nagaoka[1]

Chapter Summary

The focus of the Applied Biogeography course that I teach is on urban ecosystems and "nature" in urban contexts. The goal of the class is to demonstrate how biogeographic and ecological principles are used to explain how and why some native species in urban environments are threatened and how some non-native species have become invasive. By focusing on urban contexts, "nature" becomes an everyday experience rather than exotic, and thus students are a part of the problem and the solutions. To apply the concepts presented in class and provide evidence of proficiency, I have developed a mapping exercise and an end-of-class project. The exercise is a mapping project that documents the distribution of native trees on campus and evaluates the possible reasons for their current distribution. The class project consists of a blog and an informational brochure. For the blog, students work in groups to research an urban native or invasive species whose distribution has been affected by human actions and is in need of management. The blog contains information that corresponds with weekly class topics, which include species' life history, changes in distribution and impacts of human actions, as well as ecosystem services and current management efforts. For the brochures, students take the information developed in their blogs and individually create an informational and persuasive brochure on their species with specific action items aimed at a particular audience such as homeowners, donors, or policymakers. The products of the mapping exercise, blog, and brochure suggest that students are on the path to becoming informed citizens who can impact local environmental management practices.

1 Department of Geography and the Environment, University of North Texas

Introduction to the Course and University

The course I teach is GEOG 3240: Applied Biogeography, in which students learn basic biogeographic principles and apply them to conservation and environmental management (Richardson and Whitaker 2010; Whitaker et al. 2005). The course is housed in the Department of Geography and the Environment at the University of North Texas and contributes to our new sustainability certificate. While the course counts as a major elective for Geography students, I designed it to also attract students in other majors with "environmental" interests. The course introduces students to ecological and biogeographic principles but also covers ecosystem services and environmental philosophy, ethics, and policy.

The University of North Texas is classified as Research 01, a public doctoral institution located in the Dallas-Fort Worth Metroplex. The university is the largest in the metroplex with over 40,000 students. The student body is diverse: 53% female, 41% first-generation, 23% Hispanic, 14% African American, 1.5% Native American. The demographics of the department are not as diverse as the University (about 36% women and 25% underrepresented minorities), but diversity has increased over the last decade as the faculty has diversified.

Geography is a diverse discipline that incorporates social sciences and humanities (human geography), natural sciences (physical geography), and computer and information science (GIS and remote sensing). Most students do not know about the full breadth of the discipline. In high school, students take a social studies class or AP Human Geography. Thus, their understanding of Geography is often limited to National Geographic Bee facts about countries and landforms. College students come to Geography in two ways. First, GIS attracts students who are looking for marketable job skills. Most of these students have military backgrounds where they were exposed to GIS and spatial data on a daily basis. Students can earn a certificate in GIS or pursue a degree in Geography. Most students who become majors are socially or environmentally conscious and are seeking to make a difference. They often are recruited through our non-majors courses in human geography and physical geography that promote the relevance of geography for addressing societal and environmental problems.

I have taught the class intermittently for nine years. The initial structure for the course was more traditional, focusing on lectures, readings, and a traditional research report that required students to evaluate how biogeographic principles were used in the conservation of an endangered species. Students could choose a species from the International Union for Conservation of Nature (IUCN) Red List of Threatened Species (IUCN 2021) that was in the vulnerable, endangered, or critically endangered category. Invariably, students chose well-known exotic species such as snow leopards and pandas. I found this approach to be problematic for several reasons. It focused on species located in remote locations, thus perpetuating a human-nature dichotomy (Callicott et al. 1999). Even most endangered North American fauna are found in remote locations. Thus, nature is somewhere else and can be maintained if the species are isolated and protected from human activities. In class discussions, students' perception of "wilderness" was of an untouched nature often seen in nature documentaries. Because nature was exotic, students were often far-removed from the problems and solutions. In many instances, the only way for students to help these organisms was to donate to conservation non-profit organizations.

By switching the emphasis of the project to urban ecosystems, nature shifted from a remote, exotic, untouched ideal to habitats that students experience every day. Nature exists in their own backyards, in their apartment complexes, on campus, and in their cities. It is part of the urban landscape rather than isolated from people. Reorienting the focus of the class and the projects allowed students to understand how land use and development choices impact wildlife. The local urban context is also where they can have a direct impact as informed citizens.

The study of urban ecosystems and ecology has been gaining ground. For example, The Nature Conservancy started out as an organization similar to the National Park Service that focused on the preservation of natural areas by purchasing land to limit human access to threatened ecosystems. Recently, however, the organization has developed a mission that is more conservation-oriented (The Nature Conservancy 2021). The focus is now on conserving nature as a resource that provides services to all organisms, including us. Thus, their priorities now include providing food and water sustainably and building healthy cities.

There are also a number of relatively new journals that focus on urban ecosystems (e.g., *Landscape and Urban Planning, Environment and Planning, Urban Ecology, Urban Ecosystems*). PBS's Nature has a number of videos on urban fauna such as raccoons, foxes, squirrels, and coyotes. One online PBS video series called Urban Nature (https://interactive.wttw.com/urbannature#!/) has short 5 to 10 minute videos on what organizations are doing to mitigate the effects of human activities on species' habitats in cities such as Chicago, New York, and San Francisco. The videos are useful for showing students that nature exists in cities and there are many things that can be done to improve urban ecosystems.

In this chapter, I introduce two class projects aimed at helping students understand the biogeographic and ecological processes associated with urban ecosystems, how we impact them, and what students can do to improve them. The first is a mapping project of campus trees that is set up to correspond with topics as they are presented during the semester. The second is an informational blog and a brochure on an urban species. To set up both projects, I start with a discussion about nature in urban environments.

Students often have to be convinced that urban environments provide opportunities for "nature." Initially, students see their built landscapes as devoid of nature and wildlife or so highly modified from the original natural environment that it is impossible to go back. But I use a chapter from Michael Pollan's (1991) book, *Second Nature* to open up the possibility of nature can exist in urban environments. Pollan draws an interesting comparison that I found useful between land managed for preservation and a backyard garden. A national park or a Nature Conservancy property may be curated as an example of a wilderness, but it is still managed, albeit with the intent of supporting and preserving wildlife and the natural ecosystem. However, gardens could have a similar mission. The difference is that with gardens, the process is more about restoring ecological function than preservation. Often this is where students see the limitations of scale. Gardens, parks, and green spaces are finite pieces of land so that the impact would require many people working together toward the same restoration goal. This often seems like an impossible task to students. Then I introduce the role of policies and initiatives. How would the local environment change if city ordinances required all residential landscaping for new developments to include mainly native plants? Or what if the University developed an initiative to transform the grounds to native plants by working with local master naturalists and creating service-learning projects for students? This discussion was important for setting up both the exercise and project because it allowed students to see that the urban environment could be managed to provide more ecosystem services through individuals working together and that policy recommendations were an important part of this process.

Biogeography of Native Tree Species on Campus

Land use change and habitat fragmentation are common in urban areas. I use the local urban context to provide a setting for examining how human actions can impact the distribution of plants and animals. The Dallas-Fort Worth (DFW) Metroplex is an excellent place to study the effects of human development on the distribution of organisms across space because the growth has occurred recently. For example, in Denton, Texas, where the University of North Texas (UNT) is located, the city's population has increased from about 80,000 in 1990 to over 130,000 in 2017 (U.S. Census Bureau 2000, 2017). This rapid growth is part of a larger population trend in the Dallas-Fort Worth Metroplex, which has grown from 3.9 to 7.4 million during the same period. Associated with this increase in population was an explosion in housing developments across the region. What was formerly agricultural or undeveloped land was converted to residential and commercial land use with large expanses of impervious surface. Historical imagery on Google Earth Pro (https://www.google.com/earth/versions/#earth-pro) is very useful for showing how urban landscapes have changed. The satellite images and aerial photos go back decades depending on the location.

In Denton and the DFW Metroplex, the dominant ecosystems are prairie and oak savanna/woodland. The oak habitat is known as the Cross Timbers ecoregion, which extends from central Texas through Oklahoma and into southeast Kansas. Post oaks (*Quercus stellata*) and blackjack oaks (*Quercus marilandica*) are the cornerstone species of this ecosystem (Francaviglia 2000). Both tree species do not handle disturbance well. Post oaks, in particular, are sensitive to root disturbance. They prefer sandy soils, but in the north Texas region, these soils are shallow and over clay deposits. Since post oak roots cannot penetrate the clay, they extend laterally through the sandy soils and may extend to twice the diameter of the canopy. Thus, construction that excavates near a tree or paving that compacts the soil around the roots can result in tree mortality. Because post oak roots are sensitive to change, they are difficult to replant and propagate in a nursery setting. Thus, preservation is the main strategy for managing this species.

For the exercise, we examine the distribution of trees on the university campus as a microcosm of the city and the metroplex. UNT was first established on the periphery of an older residential neighborhood in 1891. The houses in these neighborhoods were typically built in and around the established native trees, which can be seen in historical yearbook photos and aerial photos taken in the 1940s. As the university has grown, it purchased these residential properties for new buildings. The questions that the exercise deals with are, *as UNT grew and land use changed from residential to institutional, were post oaks and blackjack oaks retained? If so, are they associated with different periods of building construction?* To answer these questions, the exercise was broken into several steps that are associated with various topics in the course.

The first step is associated with the species concept. Part of this lecture discussed the importance of identifying and organizing organisms through the Linnean system for understanding the diversity and distribution of the natural world. In the exercise, students were taught how to identify trees to species using keys. The landscape design for the Environmental Science Building, in which our department is housed, includes a high diversity of native trees. Thus, it was an excellent place to learn tree identification. We spent one class period in the classroom learning how to key out trees from leaf cuttings that I had taken from the trees around the building. It was particularly important that the students be able to identify post oaks and blackjack oaks for the next part of the exercise. For homework, the students' skills were tested. Ten trees around the building were selected that had to be identified.

The next step in the exercise was to map all trees on campus. This step coincided with the class topic on species distributions and the abiotic and biotic factors that affect where species occur. The first part of this section deals with the challenges of identifying the geographic extent of a species. Field surveys are time-consuming. Thus, species distribution modeling has become common practice (Elith and Leathwick 2009; Miller 2010). However, these models often produce the fundamental niche or where the abiotic factors are favorable for a species to live. Since these models often do not take into account the biotic factors that can affect species distribution, they may not accurately represent the realized niche. For example, all of campus is in the fundamental niche of post oaks and black jack oaks. The campus lies in the Cross Timbers ecoregion and contains the sandy soils that are commonly associated with these oak species. But people outcompete the oaks for space. Thus, the realized niche of these Cross Timbers oaks is greatly reduced. To document the distribution of these tree species on campus, students mapped and identified trees. The campus was divided into sections with a group of students responsible for mapping the trees in one section. All post oak and blackjack oak trees were identified. Any other species could be identified if the student was confident in their identification. This was optional and typically done in sections where there were no oaks. Students wanted to know what trees had replaced the oaks.

One challenge in any citizen science or student project focused on biogeographic distributions is geolocating the taxa of interest so that it can be uploaded into a GIS program. There are a few ways this can be done. Organizations often have websites where the public can report in information. For example, the Audubon Society has eBird (https://ebird.org/map/), where birders can submit a list of birds they have identified at a location. The data have been important for documenting trends in bird distributions and migration patterns (https://ebird.org/science), and are often made available for public use (e.g., eBird observation dataset, https://doi.org/10.15468/aomfnb). The precision of the locational data often varies depending on what the website records and on the data a person is willing or able to provide. A contributor may provide an address, a point location using the GPS on their phone, or zip code. The level of precision may be of concern for taxa that are less mobile. For example, a location that identifies a parcel or house lot may be precise enough for a bird species, but the location within the lot may be preferred for taxa such as insects or plant species where microenvironmental variability might be important.

Another option for recording locational information is to use a phone app. Open Data Kit (https://opendatakit.org/) is one of many free, open source software for collecting data. It is simple to design a data collection form that can be used on an Android phone. The data can include photos and geolocation. One issue is that the geolocation precision is greatly affected by the methods that are being used to determine location. Your phone may be using a combination of information from cell towers, WiFi signals, and GPS. If cellphone data is turned on, the location is determined by cell towers. The higher the density of cell towers in your area, the more precise your location information will be. If WiFi is being used, then the precision level can be about 30 m. But often WiFi is limited to indoor areas. When the GPS is used, precision is about 5 m. However, the device must be in direct line with at least three satellites. Thus, weather and overhead obstructions, such as tree canopies, affect precision. Many of these apps now provide a map within the data collection form to bypass or supplement the GPS data so that location can be manually identified. But again, the precision can vary depending on the user's ability to identify their location within an online map. Thus, phone apps can be very convenient, but geolocation data may be highly variable.

For the data collection and mapping of campus trees in this exercise, I chose to use paper maps to record the location of the trees and to take notes. Because this is a geography class, I wanted students to learn to field map and to create data layers in GIS. Doing old-school paper maps requires students to orient themselves and the object they are recording in space and in relation to the map. The process of estimating where on the map the tree is located requires spatial awareness. Students also have to record species information on the map. Thus, the paper map is a draft of what the GIS map will look like which the students are generating as they go.

The downside to paper field maps compared to a digital recording system is aggregating the individual field maps, which can be a time-consuming process. Each map has to be georeferenced, which is a multistep process within a GIS program. However, the online program, Field Papers (http://fieldpapers.org), simplifies the number of steps significantly. The program generates an atlas of paper maps for your study area that can be easily georeferenced. You zoom into your chosen study area using their online map. The program has several different base maps to choose from, including OpenStreetMap (https://openstreetmap.org), which is an open source, a community-generated map similar to Google Maps. You can then choose how many panels you want your study area divided into. The more panels, the higher the resolution of each individual map. The program will create an atlas with each panel as a separate map, each of which has a unique QR code and control points for georeferencing the panel.

To map the campus trees, a pair of students was assigned a map panel. The time spent mapping the section was about an hour depending on the number of trees in the panel, which varied from 50 to 100 trees. Since we were focusing on identifying post oaks and blackjack oaks, time spent identifying trees varied by panel. To upload the maps into a GIS program, the paper map was photographed using a phone and uploaded as a jpeg file into Field Papers. The program transformed the jpeg image into a geoTIFF file, which georeferenced the image. When the file is uploaded into a GIS program, it is automatically placed in the correct geographic location.

For the third part of the exercise, students were trained on how to use GIS during one class period. For this class, we use QGIS (https://www.qgis.org/), which is a free open source GIS program. As a geography department, we have access to ArcGIS, which is the industry standard. However, digitizing and editing is simpler in QGIS. In addition, students may not always have access to ArcGIS depending on where they become employed. ArcGIS is an expensive program so smaller businesses or non-profits cannot afford the software. Thus, by using QGIS, students gain experience working with an open source software program.

In QGIS, students work with four shapefiles or layers. The first layer is the geoTIFF file of their field map. The second layer is a new point layer or shapefile that will contain the digitized tree locations. This new layer is like a clear plastic sheet that is placed over the paper field map. The location of the trees is added to this second layer as points, and species information is added to the attribute table for each point. Since the field maps are hand-drawn, the locations of the trees are estimated. The tree locations that have been digitized need to be moved to the correct location by editing the points. The third layer students work with is a satellite image base map. The second layer with the tree locations is placed over the base map. If a tree is in the wrong location, the point is moved to the center of the canopy of the tree as seen in the satellite image.

The last step in this exercise is to analyze the distribution of post oaks and blackjack oaks in relation to development on campus. The building dataset is the fourth layer used. It contains the building footprint and the age of the building (Figure 6.1). Students submit a two- to three-page report evaluating the distribution and the potential causes for the distribution. They are asked to recall the abiotic and biotic factors that affect species distributions and evaluate which ones could be factors for the distribution of Cross Timbers species on campus. In general, the students correctly identify the most important factors. Most abiotic factors are fairly uniform across campus with the exception of water availability and solar radiation. Trees in parking lots tend to have less access to these resources than trees in a grove or those surrounded by grass or landscaping. While trees may be affected by disease and insect infestation, the main biotic factor that could affect post oak distribution is competition with people for land. Trees can be removed or preserved during construction. The post oaks and blackjack oaks tend to cluster on the north side of campus, which corresponds to the oldest

Figure 6.1. Final map showing the distribution of post oak and blackjack oaks across campus. Buildings are represented by the date of construction

part of campus with buildings built before the 1930s. However, even though preservation of the oaks seems to be associated with older buildings, it is unclear whether the pattern is due to changing development practices. The south side of campus has seen change in use of the area. A recreation complex built during the same early period as the north side of campus included the original football field, tennis courts, swimming pool, and ampitheater along with several academic buildings that were located in this area. Thus, the paucity of post oaks and blackjack oaks on the south side of campus may have been related to the earlier land use.

Students are also given access to historical and aerial photos of campus which they can use to illustrate the distribution of trees in and around campus and how that changed over time as the university grew. Early photos show buildings nestled in a grove of oaks between buildings and a surrounding treed residential neighborhood. From the aerial photos from the 1940s, the residential neighborhoods that would be replaced by university buildings all have trees. But it is difficult to tell from some of these photos the tree species. However, there are a few instances where we can identify post oaks living today in the aerial photos. Based on the distinctive canopy shape of the post oak, it is likely that many of the trees surrounding residences were post oaks. Even the south side of campus retained a large number of trees surrounding the recreational complex, including a large grove. Thus, the current distribution of oaks in south campus are due to more recent building phases.

In a follow-up session, I also use historic and aerial photos to explore how development within the city as a whole has changed. It is common practice now to clear-cut trees in most new housing developments. Thus, the post oaks are disappearing from many of these new residential areas. However, in older neighborhoods, trees were retained and homes were constructed in among the trees. These changes in construction practices provide an opportunity to examine and discuss city and campus tree policies and how they do or do not mitigate tree loss.

In the future, I plan to add in a component of evaluating ecosystem services provided by trees. The US Forest Service has a set of tools, called i-Tree (https://www.itreetools.org/), which is aimed at facilitating urban forest management. The i-Tree Eco tool calculates ecosystem services such as mitigating pollution, sequestering carbon, and reducing runoff. Students will be required to collect additional data about the trees including diameter at breast height (DBH), tree height, and canopy area. By using this tool, we can provide a client, like the city of Denton or the university, a "value" for the tree. Since a tree can provide many different ecosystem services, the students will need to target and prioritize particular services depending on the client. For example, the university would be interested in carbon sequestration to use in their greenhouse gas inventories, as well as runoff mitigation because flooding is a common problem on campus. Adding this component to the exercise will better relate the ecological and biogeographic sides to the policy and outcomes.

Another addition will be a discussion of species versus ecosystem conservation. This exercise focuses on Cross Timber species of oaks and conservation of those species. But the reality is that the Cross Timbers ecosystem is preserved only within small parks and green spaces in the city. The understory and much of the wildlife, particularly bird species, are missing. So while areas on campus retain post oaks and blackjack oaks, they no longer reflect the Cross Timbers ecosystem. But without those keystone species, ecological restoration would not be possible.

The strength of this biogeographic exercise is that it provides relevance to the lecture content of the class. And there are a number of different ways the methods can be used to study urban biogeography or ecology. The distribution of any species or taxa that might be relevant to a campus community or city, such as wildflowers, bees, or coyotes, could be studied. Habitat fragmentation becomes a bigger concern with such species. But studying invasive species can also be informative. For example, Bradford Pears (*Pyrus calleryana*) have become a common landscaping tree used in residential developments west of the Rockies, including in Denton, but it is considered an invasive species by many ecologists (Culley et al. 2011; Hardiman and Culley 2007; Vincent 2005). The tree also has a short lifespan (20–25 years) and is prone to branches splitting off the trunk during storms. In addition, non-native tree species are less likely to have the same insect biodiversity as native trees, which impacts the presence of species that feed on them (Tallamy 2007). The distribution and density of Bradford Pears can easily be documented by driving around neighborhoods in a city, especially in spring when they are in bloom. Cities with different landscaping requirements for new developments could be compared.

Components of the exercise would be relevant for any course teaching human impacts on or relationships with plants and animals because most of these processes have a spatial or geographic component to them. Mapping, in particular, provides a means to examine relationships between variables across space. The biggest challenge of using this exercise is teaching basic GIS skills. But there are many tutorial videos online for novices. Even though this is a Geography course, only half of the students had taken a GIS class, but students were able to pick up the skills quickly. Mapping is becoming more accessible and a more common graphical component to analysis. For example, a map can be generated in Excel using the PowerMap add-in. Also, any mapping project could be done using old-school techniques. A poster-sized map of the study area that students plot their data on using push pins, stickers, markers, or post-its could be used. If hung in a public area, the progress of the class project could be open for everyone to see.

Informational Blog and Brochure on Local Native and Invasive Species

The final project for my course used to be a traditional research report as a means of evaluating how well students had digested the course content and could apply it to a species. The paper required students to evaluate how biogeographic principles were used in the conservation of an endangered species. Students could choose a species from the IUCN Red List (IUCN 2021) that was in the vulnerable, endangered, or critically endangered category. However, as mentioned above, the focus on endangered species perpetuated the idea that nature is remotely located somewhere else and is not a part of urban environments. The paper format was also uninspiring because students did not receive feedback on their papers since it was an end-of-semester project. The course content also could have been evaluated through exams, but this seems to focus students' attention on memorizing facts rather than understanding processes.

As an alternative to exams and final papers, I decided to evaluate if students have gained an understanding of biogeographic concepts presented in class through a group project using blogs and a brochure that is submitted individually. The students choose a local native species that is affected by urban development and or an invasive species that is having an impact on the native landscape. In Texas, the native species have included coyotes, native bees, and post oaks, while invasives include introduced taxa such as wild pigs, zebra and quagga mussels, and landscaping plants like Chinese privet.

For the blog, students work in groups of three to four to develop the content. The goal is to apply the concepts covered in class. Thus, all of the weekly topics—spatial distribution, range reduction/expansion, life history factors, the impact of habitat loss and fragmentation, ecosystem services, and mitigation practices—need to be applied to the species of interest. Students can develop their blog through free websites or through the course learning management system. The blogs are evaluated on whether all the biogeographic concepts covered in the course are applied to the species' context. For example, students studying a native species have to demonstrate a change in the distribution of the species, as well as identify the relevant abiotic or biotic factors affected by urban development that impact the distribution.

The blog format provides a few things that are not possible with a more traditional class paper or presentation. One aspect that differs is that each group develops a near-final draft that the other students then provide feedback on. Students are given a week to comment on the six to seven blogs. They are not allowed to ask the same questions or comments, but they can build off comments already made. The comments on the blog are graded as participation points. I found the student comments to generally be useful and insightful and it has not been necessary to monitor the quality of the comments. Students seem to be engaged with the evaluation process for a couple of reasons. As a consumer of the information in the blog, students are trying to understand the issues associated with each species. They have gone through a similar process researching their own species so they have expectations about what should be covered in the blog. And because the students are going through a similar learning process with their own species, reading other groups' blogs helps students see what they missed or could expand on in their blogs. After the student comment period ends, I read the blogs and the comments and provide direction to each group on what they need to address in their blogs. The groups then take those comments and submit their final blog for grading with the new information highlighted so I can easily identify what has been edited and added. Not surprisingly, the information in the final blogs is much more thorough than what I typically used to see in end-of-semester papers.

The blogs are an effective means of getting students to assimilate a lot of information about their species so that they can better understand the management issues. Students then take the information from the blogs and reorient it into an individually written persuasive brochure related to species management. The students have to make several choices for the brochure: identify the audience that they want to target, define the role they are playing, identify the goals that they are trying to meet, and determine the information or data that should be used to support those goals. For students working with native species, the brochures are typically oriented toward getting the general public to provide or not remove habitat. For the invasive species, the brochures raise awareness about the problems with these organisms in urban environments. Students also have to submit a short (one-page or less) statement as to why they thought it was important to focus their brochure on the particular issue and how what they were advocating for could affect the species' distribution and abundance. The brochure, as an additional product to the blog, provides a few different things. It allows students to be creative with their knowledge. Students in the same group often came up with completely different approaches to the brochure. And most of the brochures were well-laid out, with good graphics and short concise but effective text. Since students work on these brochures individually, it also became clear who the freeloaders were in the group project. Those students who researched their species thoroughly developed brochures that had more depth to them.

In one brochure, a student was studying coyotes and chose to focus their brochure on minimizing human-coyote conflicts. They took on the perspective of animal services in the city, and their audience was homeowners. Coyote sightings are not uncommon in Denton, even in the middle of town. There are still avenues of green spaces connected with riparian zones that provide denning and allow for coyotes to move extensively across the city. Every year, there seems to be the local news story of missing pets due to coyotes or bobcats. The student was concerned that some homeowners would take more drastic actions such as shooting or poisoning coyotes and wanted to allay their fears about coyotes. The brochure presented basic information on coyotes including the benefits of having coyotes in the city, and provided actions that the homeowners could take to minimize pet predation (e.g., bringing pets in at night, not leaving food out, etc.).

The brochures are graded on having clear roles for the author and the audience. They also have to have a clear goal and effectively and persuasively present information to meet that goal. I was pleasantly surprised at how many of the students were able to create effective targeted persuasive brochures. This product was also important for helping students understand different viewpoints of various stakeholders and communicate concrete solutions. The blog and brochure could be used in almost any course that is trying to help students process course content, apply that content to a particular context, and develop solutions.

Discussion and Conclusions

As most instructors know, there are challenges to implementing any exercise. For example, I regularly have group work projects even though they are often shunned by students because of the unpredictably of working with other students. In both exercises presented here, the group work is mainly done to gather data and information. Using groups during this phase of the exercises is useful because the data collected is generally more accurate with multiple people collecting and checking data. In addition, this part of the exercise is graded qualitatively (e.g., full, partial, no credit) on the information collected in the blog and the tree data collected. The final grade for the projects is based on the individual final products. It is obvious from the quality of the final products which students fully participated in the activities. Thus, freeloading is not a useful strategy for these exercises and it alleviates concerns by fully participating students that others will benefit from their work. Another option for mitigating freeloading is to use group member evaluations in the grading process. However, I found that students are often unwilling to be responsible for accurately rating another student's performance. They do not want to be the ones who decide that someone else should receive a lower grade even when a student does not contribute as much to the project.

One option that has worked well for dealing with freeloaders is mediation and possible expulsion. Students can file a complaint about a group member's participation in the project (e.g., unresponsive to emails, does not attend meetings, does not complete tasks on time, etc.). I then reach out to the student about the complaint, provide suggestions to minimize the problems, and issue a warning that if the issue is not resolved, the group will be able to "vote them off the island." In such a case, that student would then have to complete the assignment on their own. In the 20 years that I have been using this option on group assignments, I have rarely had to remove a student from a group. Instead, students are more likely to ask for help to work out their grievances rather than filing a complaint. Or when a grievance is filed, the contact with the instructor indicates to the underperforming student that someone is paying attention to their actions, which is usually enough to motivate them to change how they interact with their group members.

These exercises would also need to be modified for different class types. Large lecture classes would have to modify the exercises to deal with the logistics of managing and training students and then grading student work. The campus biogeography exercise may be easier to do with a larger class because there would be more data collectors involved. Students could generate data as a class, then evaluate the data as part of a short answer or multiple question test. The exercises could also be turned into service-learning projects if partnered with an office on campus (e.g., sustainability, grounds) or local groups off-campus.

Since these exercises examine issues specific to our campus, they would have to be modified to fit the local context. However, the strength of a (bio)geographic perspective is that it is scalable and integrative. The exercises can focus on the distribution of species, ecosystems, or environments in a neighborhood, on campus, or within a city. They can be integrative in that a variety of environmental factors are examined, both natural (e.g., hydrology, soils, etc.) and built (e.g., impervious surface, heat island, etc.). This provides a lot of flexibility to modify the exercises to fit different contexts. For example, the campus exercise, in particular, may need to be modified from improving the distribution of native species to enhancing the built environment with better managed natural landscapes. If the campus has been highly developed with few to no native species remaining, then the research can focus on the distribution of green space or tree canopy and their impact on runoff, urban heat island effect, recreational opportunities, or other ecosystem services. It could be coupled with brief face-to-face surveys of students about their perceptions of green space and/or grounds and maintenance workers about the challenges of maintaining or improving the campus landscape. Having used this strategy in other courses, I have found that these types of surveys provide a sample of viewpoints from different stakeholders that broadens the perspectives on the issue for the students in the class. Another strategy would be to scale up and evaluate how the campus fits in with or is affected by the surrounding environment. If the campus has little green space but is surrounded by undeveloped or underdeveloped land, then it may be contributing to habitat fragmentation. If the opposite is true, then the campus may be a habitat island. Thus, the campus exercise can be modified to document the opportunities to integrate campus into the surrounding area through corridors or establishing habitat.

In sum, the tree biogeography project and the urban species blog and brochure have been an effective means in my course for teaching students several things. Both activities require students to apply the concepts discussed in class and expose students to "nature" in urban contexts. They also provide a means for helping students understand how human activities affect the distribution of wildlife and encourages them to identify solutions. Most of the students who take my class are inherently interested in environmental issues and are looking for ways to make a difference. Structuring projects that are based on campus or that focus on urban species and ecosystems seems to provide a means for students to better identify how they can improve their local environments. In addition, they learn that many of the issues facing native urban wildlife can be mitigated by an active and informed citizenry.

References Cited

Callicott, J.B., L.B. Crowder, and K. Mumford. 1999. Current Normative Concepts in Conservation. *Conservation Biology* 13:22–35.

Culley, T.M., N.A. Hardiman, and J. Hawks. 2011. The Role of Horticulture in Plant Invasions: How Grafting in Cultivars of Callery Pear (*Pyrus calleryana*) Can Facilitate Spread into Natural Areas. *Biological Invasions* 13:739–746.

Elith, J., and J.R. Leathwick. 2009. Species Distribution Models: Ecological Explanation and Prediction Across Space and Time. *Annual Review of Ecology, Evolution, and Systematics* 40:677–697.

Francaviglia, R.V. 2000. *The Cast Iron Forest: A Natural and Cultural History of the North American Cross Timbers.* University of Texas Press, Austin, TX.

Hardiman, N.A., and T.M. Culley. 2007. The Beginning of a New Invasive Plant: A History of the Ornamental Callery Pear in the United States. *BioScience* 57:956–964.

International Union for Conservation of Nature (IUCN). 2021. The IUCN Red List of Threatened Species. Version 2021-1 [web page]. URL: https://www.iucnredlist.org. Accessed on August 11, 2021.

Miller, J. 2010. Species Distribution Modeling. *Geography Compass* 4:490–509.

Nature Conservancy, The. 2021. About Us: Who We Are [web page]. URL: https://www.nature.org/en-us/about-us/who-we-are/our-history/. Accessed on August 11, 2021.

Pollan, M. 1991. *Second Nature: A Gardener's Education.* Grove Press, New York, NY.

Richardson, D.M., and R.J. Whittaker. 2010. Conservation Biogeography—Foundations, Concepts and Challenges. *Diversity and Distributions* 16:313–320.

Tallamy, D.W. 2009. *Bringing Home Nature: How You Can Sustain Wildlife with Native Plants.* Timber Press, Portland, OR.

Vincent, M.A. 2005. On the Spread and Current Distribution of *Pyrus calleryana* in the United States. *Castanea* 70:20–31.

Whittaker, R.J., M.B. Araújo, P. Jepson, R.J. Ladle, J.E.M. Watson, and K.J. Willis. 2005. Conservation Biogeography: Assessment and Prospect. *Diversity and Distributions* 11:3–23.

U.S. Census Bureau. 2000. County and City Data Book: 2000 [web page]. URL: http://factfinder2.census.gov. Accessed on February 20, 2019.

U.S. Census Bureau. 2017. American Community Survey Annual Estimates of Resident Population Change for Incorporated Places Over 50,000 or More in 2017 [web page]. URL: http://factfinder2.census.gov. Accessed February 20, 2019.

Chapter 7.

Getting into the Weeds with Experiential Learning: Discovering Where Medicinal Plants Grow

John Richard Stepp[1]

Chapter Summary

This is an exercise based on a teaching experiment first developed at the University of Georgia and now taught at the University of Florida. In its current form, students are divided into two or more groups and assigned either anthropogenic ("disturbed") habitats or forested habitats. They collect or identify every species within a pre-defined unit and then consult literature and field guides to determine if the plants within the area are medicinal and/or weedy. Students are often surprised to learn that disturbed habitats usually contain a much larger number of medicinal species. The learning objectives are to evaluate different types of habitats and discern which are potentially most useful. One alternative approach to the exercise is to instead focus on wild edible species. The results are similar and the alternate focus allows for a robust conversation and discussion about the nexus and overlap between food plants and medicinal plants.

Introduction and Justification for the Exercise

There is a common perception among the general public that forests (especially tropical forests) are the most significant habitats for medicinal plants and, by extension, places to discover new pharmaceuticals (Stepp 2004, 2018; Stepp and Moerman 2001; Voeks 2018). This assumption was heavily promoted by environmentalists starting in the 1980s as a reason

1 University of Florida, Department of Anthropology

to conserve forests—it provided a strong economic justification for conservation while also holding out the promise of a "magic bullet" to cure some of the world's worst diseases, both old and new. It is perhaps no coincidence that this time period also saw the widespread emergence of a new and deadly infectious disease (AIDS) that, at the time, was seen as particularly complex to treat and virtually uncurable. Hollywood also grabbed onto to this trope with films such as Medicine Man (McTiernan 1992) starring Sean Connery, and the mainstream media featured numerous news reports of the promising medical cures that were waiting to be discovered in the rainforest. Ethnobotanists were not immune to making this argument as well, and for obvious reasons: It brought new found interest and prominence to a relatively obscure field and led to research funding from governmental organizations, foundations, and the private sector.

While these sort of "save the rainforest for the cures" claims were most prevalent in the 1980s and 1990s, the argument persists to this day. A Google search yields numerous sites making these assertions, often with highly dubious statistics. For example, the Medicine Hunter website (Medicine Hunter Inc. 2021) notes that "the Amazon rainforest contains about 5 million species of healing plants, with more being discovered every day!," a claim made all the more remarkable considering that a liberal estimate of the flora (vascular plants) of the entire world consists of 391,000 species (Royal Botanic Gardens, Kew 2016). The website of the NGO Rainforest Trust (https://www.rainforesttrust.org/owed-to-nature-medicines-from-tropical-forests/) proclaims "Do We Know What We are Losing? The answer is NO!" in an article entitled "Owed to Nature: Medicines from the Tropical Forests" (Torrence 2013). While there are clearly medicinal plants found in the tropical forests, the narrow focus on forests and trees obscures the reality that there are significant medicinal plants that do not come from forests and instead thrive in anthropogenic habitats.

A common logical fallacy in these sorts of articles is to list a few major pharmaceuticals derived from plants (with the Madagascar rosy periwinkle, *Catharanthus roseus,* often with top billing) followed by a mention of the large number of plant species found in tropical forests. Using the rosy periwinkle as the poster child for drugs derived from rainforest plants is particularly problematic because, while it is a plant of tropical origin and could be found in tropical forests, it is also a weed with global distribution and thrives in all sorts of habitats, especially human modified ones. This corresponds with a finding from the early 2000s that of the 95 plant species used at the time for prescription drugs, 45 were considered weeds (Stepp 2004). While the number of pharmaceuticals derived from plant species is now higher, weeds are still significantly represented beyond what one would expect if the relationship was random.

If we add ethnobotany to the equation, we find that humans most often obtain medicinal plants in human modified habitats even when primary forest is present: in the tropics, in temperate regions, and pretty much everywhere in the world. A few notable examples include Voeks' (1997) work with Candomblé healers in Atlantic Coastal Forests of Brazil; Posey's (1984) work with the Kayapo in the Brazilian Amazon that led to the term "anything but abandoned" fields because of the number of medicinal and food plant species found in them and extensive use of disturbed habitat in general for medicinal plant procurement; Frei et al.'s (2000) findings that 80% of medicinal plants occur in non-forested disturbed area; Stepp and Moerman's (2001) comparative work between Highland Chiapas and North America that found in both regions, weeds appear with a significantly higher frequency ($p < 0.00001$) in the pharmacopeias than would be expected; along with other studies too numerous to cover here. Readers may be interested in Voeks (2018) and Stepp (2018) for a more detailed discussion.

Clearly there is a wide gulf between the general public's perception and the reality of where traditional and small-scale societies actually procure medicinal plants. This exercise provides a fun and memorable way to explore the hypothesis that human modified habitats are more significant for medicinal plant procurement than forested habitats and it also allows students to learn about ethnobotanical knowledge that exists in their own communities and regions.

The original exercise was developed for an ethnobotany course that I taught while a graduate student at the University of Georgia (UGA). It was somewhat opportunistic in origin due to the presence of abandoned industrial activity in and around the UGA campus leading to various fields and plots at different successional stages. Most notably were a series of abandoned railroad tracks a short walk from the classroom that converged on downtown and the adjacent campus. These provided a transect of sorts along which to sample successional vegetation. Originally, students collected all of the weedy species they could find growing along railroad tracks and discovered that 38 out of 60 were considered medicinal. The exercise met with great success and ended up being the subject of a human-interest story by Charles Osgood on CBS News Radio after some local newspaper coverage drew attention to the experiment.

Introduction to the Institution and Course

The exercise is currently taught at the University of Florida (UF). UF is the flagship public university in the State of Florida and remains true to its land grant institution status through a wide range of community outreach and extension. UF has agricultural (and in some cases, aquacultural) offices in all 67 counties in Florida. The university is: large, enrolling 57,806 students in the academic year 2020–2021, comprehensive (it has the largest number of different undergraduate majors in the country), and is ranked number six in 2020 U.S. News and World Report rankings of public universities. Undergraduate students come to UF highly prepared for university studies with an average weighted GPA of 4.4 and average SAT scores of 1364. There are over 6000 international students in residence.

I have taught this exercise in a number of different courses at the undergraduate level, but sometimes at the graduate level or even combined undergraduate/graduate offerings. It is most commonly taught in a course in ethnobotany, but I have included it either in its entirety or portions of it in related courses including Ethnobiology, Ethnoecology, Biocultural Diversity, Tropical Gardens, and Methods in Ethnoecology. The latter course is also occasionally taught to other academics post-PhD through a "train the trainers" pedagogy that was initially developed with funding by the National Science Foundation and held at the Duke University Marine Lab in Beaufort, South Carolina between 2006–2012. Many of the graduate courses taught at UF that are mentioned above are core courses in the Tropical Conservation and Development (TCD) program. TCD is a highly applied interdisciplinary value-added graduate certificate program with a mission to bridge theory and practice to advance biodiversity conservation, sustainable resource use, and human well-being in the tropics (Kainer et al. 2006). The program consists of over 100 faculty affiliates in 25 different departments across campus representing both biophysical sciences and social sciences. Now in its third decade, the program has trained thousands of students and counts among its alumni many leading governmental and non-governmental leaders devoted to conservation in dozens of countries across the globe.

Current Biophysical Environment Where the Activity Takes Place

The exercise has been taught in a number of locations but the most ideal one has been the UF Natural Areas Teaching Laboratory (NATL). NATL is 24 ha (60 acres) of forested and successional habitats divided into two contiguous tracts located on the southwest edge of campus behind an arts complex and the Florida Museum of Natural History. Within NATL there are three significant upland plant communities commonly found in Florida including mesic hammock, upland pine, and old-field succession at various stages (http://natl.ifas.ufl.edu). The successional stages are regularly maintained in rotation of different ages with some plots maintained at 1–2 years while others are at 10 years up to 40 years (Figure 7.1).

This area gained unfortunate infamy in 1990 for housing the campsite of a notorious serial killer nearby. Perhaps owing to the university's understandable desire to move beyond this horrific event, and coupled with the need for more land on a rapidly growing campus, the site was eyed for development. Fortunately, some prominent ecology, entomology, and biology professors saw the utility of the land for experiential learning and formed a planning group to bring awareness of the site's potential to the university community. They were successful in their efforts and by March 2001 were able to have the site formally recognized by the university as an area for experiential learning. In recent years, the oversight committee has designed trails for access along with useful signage that denotes various natural aspects of the area, common invasive species, most significant species of flora and fauna, and Lepidoptera biology amongst others (Figure 7.2). The land has been successfully used for a number of teaching experiments and natural history demonstrations. It is also regularly used for recreational

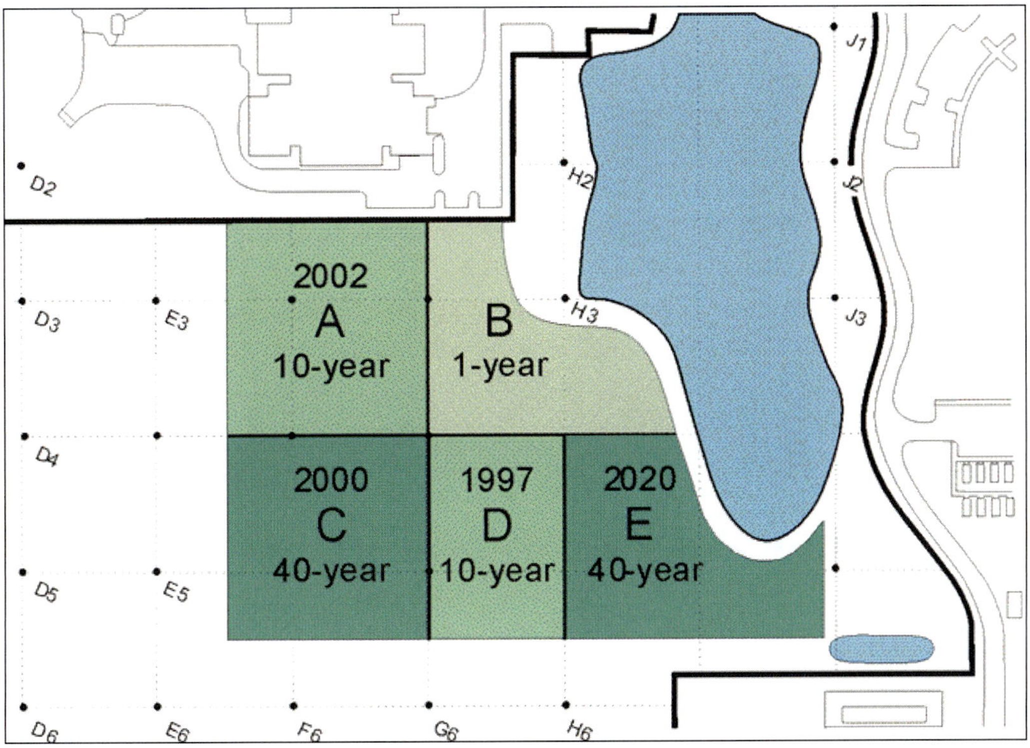

Figure 7.1. Successional stages at the UF Natural Areas Teaching Laboratory.

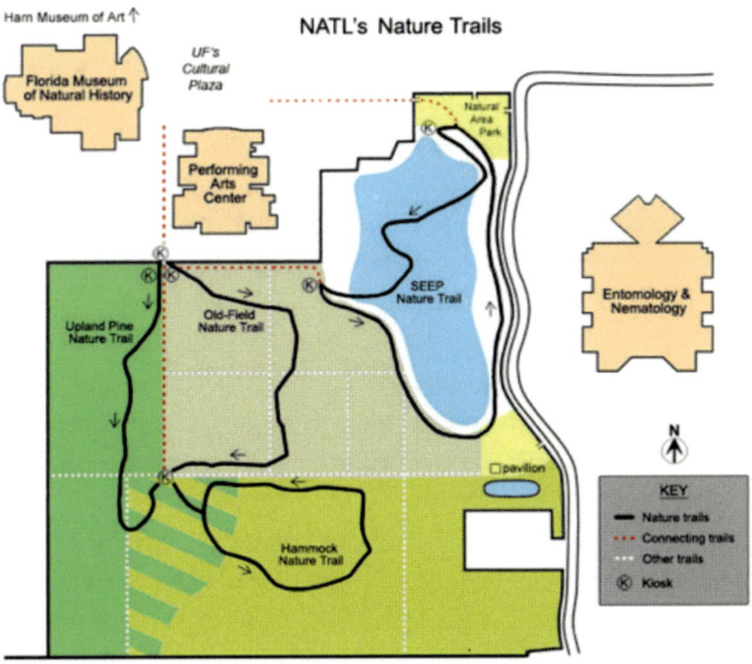

Figure 7.2. Nature trails at the UF Natural Areas Teaching Laboratory.

purposes by students and the greater surrounding urban community. The site is visited by thousands of people monthly. The presence of both forested areas as well as clearly delineated patches of different successional stages makes it ideal for this exercise.

Description of the Learning Activity and Learning Objectives

In its current form, students are divided in two (or more) groups and each are assigned either anthropogenic ("disturbed") habitat or forested habitat. The time involved can vary considerably depending on the size of plots and the diversity present. It could conceivably be done over two 50-minute course periods—one for the inventory and one for the analysis. However, I usually teach it over a three-hour block, along with one classroom hour devoted before the exercise to explain basic techniques. They are tasked with first identifying all the plant species, including trees, found within their plot of land and then determining whether the plant species are considered medicinal. The species inventory can be a relatively simple task or can take on varying levels of complexity. Much depends on the learning objectives and level of student ability and interest. In its simplest form, students are provided a relatively small plot (10 x 10 m) or short transect for which the vegetation is already well known and students are heavily supervised in their plant identification. Photos are taken on camera phones to compare back to known specimens in the classroom using either internet databases or readily available field guides. More sophisticated implementations of the exercise involve larger plots and tutorials on various types of ecological sampling. Prior to the exercise, students are instructed how to collect and prepare botanical voucher specimens and how to key out species and compare them to known voucher specimens in an herbarium. In any case, it is important to keep the sampling strategy in both types of vegetation the same size, whether it involves plots or transects. Also, it is helpful to make sure that the most botanically knowledgeable students are evenly distributed among the groups. This ensures that each group has an "expert" and also allows for students to teach each other how to identify plants and conduct field botany.

The next part of the exercise involves comparing the now known inventory of plant species and determining the medicinal status of each one. This is done in class with all the students, although instructors may want to assign this portion as homework. There are a number of internet databases that can be used in this phase. The most significant one for North America is Daniel Moerman's Native American Ethnobotany Database housed at the Botanical Research Insititute of Texas (http://naeb.brit.org). The site is easily queried and contains voluminous data for hundreds of Native American cultures compiled over several decades by Moerman. Similar databases exist for other parts of the world and the instructors will want to find an appropriate one depending on where they are conducting the exercise. It is also useful to have a number of local or regional field guides available for students to consult. Our community is fortunate to have a semi-annual used book sales sponsored by the library district. At this sale, it is common to find several copies of local plant guides for less than a dollar. I have been able to mine this sale over the years and now have a large box of field guides at the ready for this exercise.

Once the data are compiled, students compare results between the groups. The final deliverable is a spreadsheet of the botanical inventory along with notations of their medicinal status. An optional addition could be a medicinal plant guide to the area studied. After conducting the exercise dozens of times in different locations, I have never had results where the forested habitat group ends up with more medicinal plant species than the anthropogenic habitat group. Students are often surprised at this finding and a robust discussion ensues as to why this might be the case. Instructors should ensure that adequate time is provided for this discussion as it can easily last an hour or more if properly guided. I often follow this exercise with units on biochemical ecology that provide a theoretical basis for the results. I also emphasize aspects of human ecology and what it means for a human society to forage for medicinal plants, both the constraints and possibilities. The result is that students learn about the importance of disturbed habitats for wild medicinal plant resources. They also gain an understanding of ecological succession along with a number of research techniques employed by botanists, ecologists, and ethnobiologists.

An alternative form of this exercise is to focus on wild edible plants. This can be done with the existing datasets collected, or the primary focus at the onset of the exercise can be food plants. Regardless, the results are often the same—with anthropogenic habitats containing more edible species. The subsequent discussion can follow much of the same lines of reasoning as the medicinal plant exercise but with a focus on human nutrition. The instructor can also introduce the fact that a continuum exists between medicinal and edible plants and many important plants are considered to be both. An added advantage of a focus on wild edibles is that the exercise can end with a wild foraged food dinner. However, I strongly encourage sharing with the forested habitats group so they don't go hungry!

Reflections and Conclusions

This is a fun and dynamic exercise that effectively illustrates important findings and theoretical developments in ethnobotany, human ecology, and biochemical ecology. It also helps to build small group cohesion in a class, increases awareness of the biophysical environment, and leads to reflections on how small scale and traditional societies are highly reliant on their local environments for survival. It is a case study in the oft-repeated phrase, "ethnobotany is the science of survival" (Prance 2007).

The core exercise requires very little in terms of materials other than local or regional field guides to the medicinal (and/or edible) flora. It may be challenging at densely urban institutions to find suitable forested areas or disturbed areas where weeds are left unchecked and not brought into submission by overly fanatical narrow aesthetics of which plants belong in particular places. On the other hand, it is not uncommon to find temporarily abandoned vacant plots of land even in the most gentrified cities today. Having an area with a controlled mosaic of different plots at various successional stages like we have at the University of Florida Natural Areas Teaching Laboratory is not required. However, it does allow for a more fine-grained analysis when students can sample vegetation at different stages. It also helps to teach this exercise during a time of year when plants are actively growing. For instructors in the high temperate latitudes and beyond, it might prove difficult to implement during semesters filled with cold months. In general though, it is a highly flexible activity that can be modified in many different ways while still meeting the main learning objectives.

Another aspect to consider is that, based on my previous experiences, the general public is fascinated with the concept, exercise, and results, and media outlets are eager to showcase this sort of experiential learning. Instructors may want to notify their campus publicity outreach office and local or student media outlets about the possibility of developing a human-interest type story on the activity. It can also be a particularly effective way to highlight the importance of outdoor teaching spaces/laboratories.

I've spoken elsewhere of a living pharmacy (Stepp 2018), whereby people are able to find effective treatments for illnesses within a short distance from their household. This is a common occurrence in many parts of the world that have an intact ethnobotanical tradition and people are self-reliant for their primary healthcare. This exercise has the potential to bring this living pharmacy to life for students and bring a deeper understanding of the importance of all types of plant diversity and the complex connections between humans and their biophysical environments.

References Cited

Frei, B., O. Sticher, and M. Heinrich. 2000. Zapotec and Mixe Use of Tropical Habitats for Securing Medicinal Plants in Mexico. *Economic Botany* 54:73–81.

Kainer, K., M. Schmink, H. Covert, J.R. Stepp, E. Bruna, and J. Dain. 2006. A Framework for Graduate Education for Tropical Conservation and Development. *Conservation Biology.* 20(1):3–13.

McTiernan, J., dir. 1992. Medicine Man [Film]. Hollywood Picutres, Burbank, CA and Cinergi Pictures, Santa Monica, CA.

Medicine Hunter, Inc. 2021. Medicine Hunter [web page]. URL: http://www.medicinehunter.com. Accessed on December 31, 2020.

Moerman, D. 2003. Native American Ethnobotany: A Database of Foods, Drugs, Dyes and Fibers of Native American Peoples, Derived from Plants [web page]. URL: http://naeb.brit.org./ Accessed on December 31, 2020.

Posey, D.A. 1984. A Preliminary Report on Diversified Management of Tropical Forest by the Kayapó Indians of the Brazilian Amazon. In *Ethnobotany in the Neotropics,* edited by G.T. Prance and J.A. Kallunki, pp. 112–126. New York Botanical Garden, New York, NY.

Prance, G.T. 2007. Ethnobotany, the Science of Survival: A Declaration from Kaua'i. *Economic Botany.* 61:1–2.

Royal Botanic Gardens, Kew. 2016. The State of the World's Plants Report – 2016 [web page]. URL: https://stateoftheworldsplants.org/2016/. Accessed on December 31, 2020.

Stepp. J.R. 2004. The Role of Weeds as Sources of Pharmaceuticals. *Journal of Ethnopharmacology* 92:163–66.

Stepp, J.R. 2018. *Ethnoecology and Medicinal Plants of the Highland Maya.* Springer Nature, New York, NY.

Stepp, J.R., and D.E. Moerman. 2001. The Importance of Weeds in Ethnopharmacology. *Journal of Ethnopharmacology* 75:19–23.

Torrence, P. 2013. Owed to Nature: Medicines from Tropical Forests. Rainforest Trust [web page]. URL: https://www.rainforesttrust.org/owed-to-nature-medicines-from-tropical-forests/. Accessed on December 31, 2020.

Voeks, R. 1997. *Sacred Leaves of Candomble: African Magic, Medicine and Religion.* University of Texas Press, Austin, TX.

Voeks, R. 2018. *The Ethnobotany of Eden: Rethinking the Jungle Medicine Narrative.* University of Chicago Press, Chicago, IL.

Chapter 8.

Culture, Ecology, and Health in Mesoamerica: An Online Lesson

Amanda M. Thiel[1] and Marsha B. Quinlan[1]

Chapter Summary

Online learning is increasing in popularity, and human ecology and ethnobiology represent disciplines that have not yet been widely available in this unique format. We foresee this changing soon and see many potential benefits and challenges to teaching human ecology and ethnobiology online. Here, we share our experience of developing and offering the online, undergraduate lesson *Culture, Ecology, and Health in Mesoamerica*. The lesson describes and links the culture and ecology in Mesoamerica to the study and practice of traditional medicine, and explores ways for students to engage in finding creative solutions to issues of traditional ecological knowledge loss and culturally appropriate medical care. In this chapter, we reflect on pedagogical theories relevant to online learning and teaching in hopes of drawing critical attention to factors that influence student success and satisfaction. Based on our experience, we offer suggestions for developing lessons on human ecology and ethnobiology in the online environment.

The Institution and Course

Our lesson, *Culture, Ecology, and Health in Mesoamerica*, is offered within two social science courses at Washington State University (WSU). Based in Pullman, Washington, WSU has several other campuses as well as a "Global Campus" for online education. WSU is a public,

1 Washington State University

land-grant university that enrolls ~30,000 students in undergraduate, graduate, and professional degree programs.

Amanda Thiel (author) originally developed the online *Culture, Ecology, and Health in Mesoamerica* lesson that we discuss here within a graduate course, *Preparing to Teach Online*. She designed and tested the lesson for WSU Global Campus's 400-level anthropology course, *Human Issues in International Development*. Thiel later adapted the lesson as a case-study to fit the focus and objectives of Quinlan's (author) 300-level Honors College course.

Global Issues in Social Science: Human Health and Healing across Cultures (Honors 370) is a three-credit course that is one of a trio of international, research-oriented core requirements that (with Global Issues in the Sciences and Global Issues in the Arts and Humanities) offers knowledge breadth and interdisciplinary collaborative mindsets to WSU Honors College students of all majors. It is a 25-student course that uses a case-study framework to examine the dynamic forces that shape human behavior. Marsha Quinlan chose her section's topic (and course subtitle) *Human Health and Healing across Cultures*, to compliment her interests in ecological medical anthropology and ethnobiology.

Quinlan invited Thiel to test the *Culture, Ecology, and Health in Mesoamerica* lesson in this second course. Thiel shifted the focus of the lesson content from international development programs relating to midwifery, toward general ethnopharmacological studies in the region (although international development still plays a role in the lesson's assignment, as you will see). We discuss the modified, ethnopharmacological version of the lesson here.

The Lesson

Objectives

The lesson *Culture, Ecology, and Health in Mesoamerica* introduces students to the culture and ecology of Mexico and Central America, and links those topics to traditional medicine, particularly ethnopharmacology. Concerning the process of student learning, the lesson aims to engage students in collaborative and reflective methods of inquiry to not only demonstrate knowledge but also think critically and communicate effectively about this knowledge. Table 8.1 shows the objectives or desired outcomes for the students, the learning resources and activities within the lesson that support the achievement of that outcome, and the method of assessment for the achievement of each objective.

Narrative

We organized the lesson *Culture, Ecology, and Health in Mesoamerica* into an introduction, written lessons (and extra reading), a quiz, an essay with peer review, and final discussion. Each item appears in the online classroom platform as a separate page or file, and students

Table 8.1. Lesson design matrix outlining objectives and the methods used to meet them.

Goal	Objective	Resource*	Activity	Assessment
1	Demonstrate knowledge of basic cultural history and ecology of Mesoamerica	*Lesson 1: Culture and Ecology in Mesoamerica* on culture, history, and ecology	Watch linked videos that support reading	Multiple choice and short-answer quiz
2	Link and explain the importance of the study of local culture and ecology to traditional healthcare	*Lesson 2: Health and Plants among the Maya of Guatemala*, plus a relevant extra article (Michel et al. 2007)	Write a report and share on the class's discussion board	Written report
3	Design a small development project for traditional healthcare that respects and is informed by local culture and ecology	Peer review in the discussion board	Collaborate through peer review on discussion board posts	Peer response and constructive critique

*All resources are provided in Appendices 4–6.

can easily move between them to select an item to view, except for the quiz which students may only open once. The lesson takes approximately 3–4 hours to complete and students are required to complete it on their own time during one particular week of class.

The introduction has two parts. The first is a two-paragraph overview of the lesson content, learning objectives, and methods of assessment. The second part is a "to-do" list with student tasks listed as follows:

Your "To Do" List

Readings:

- Read the notes: "Culture and Ecology of Mesoamerica"
- Read the notes: "Health and Plants among the Maya of Guatemala"
- Read the article: *Medical Potential of Plants Used by the Q'eqchi' Maya of Livingston Guatemala for the Treatment of Women's Health Complaints*, by Michel et al.

Activities:

- Complete the quiz.
- Watch this 2:48-minute video, Medicinal Plant Tour in San Juan La Laguna: https://www.youtube.com/watch?v=lvtGE7MwIpU.
- Complete the writing and discussion assignment.

There are two written lessons with extra reading. Lesson 1 is called "Culture and Ecology of Mesoamerica." This lesson is more or less equivalent to a short lecture (~1700 words). We first define Mesoamerica for the course (definitions often differ), including a short discussion of geography. Next, we describe salient features of the ecology (e.g., land use and subsistence) and culture (e.g., language, ethnicity, religion). We embed maps and photos that relate to and illustrate lesson content. We also include a link[1] to a short (3-minute) video called "From Milpa to Mesa" (Perennial Plate 2016) that shows the process of food production and preparation of common staple foods in Mesoamerica.

Next in Lesson 1 is a section on history. We briefly mention pre-history and provide links[2, 3] for optional further study. The rest of the history section focuses on two intertwined themes: post-colonization patterns of unequal land distribution, and the role of foreign governments in shaping local politics. We end the section with two current issues in Mesoamerica: immigration and drug trafficking. We embed a link[4] to an interactive graph showing rates of immigration by country so that students can explore current statistics according to their interest. We also embed a link[5] to a map showing present-day drug trafficking flows in Mexico. We conclude the first lesson with three questions (below) for reflecting upon the main points and the overall relevance of the lesson content.

Questions for reflection:

1. What are some of the cultural features (language, subsistence, etc.) that constitute daily life in Mesoamerica?
2. How has the role of outsiders (foreign governments) historically affected and been received by Mesoamericans? Why?
3. What are some of the assets and challenges of Mesoamerican people, historically and currently?

We advise students to reflect on the above questions in preparation for the quiz. Students may complete the quiz at any time, and it has eight multiple choice questions and two short-answer ones. The material covered in the quiz comes from Lesson 1, so a natural time for students to complete the quiz is after that reading. The quiz is open-book, timed (60 minutes), and only available once.

Lesson 2 discusses "Health and Plants among the Maya of Guatemala." This section begins with an overview of Guatemalan ecology, culture, and health statistics. The lesson then focuses on ethnomedicine and ethnopharmacology within Maya communities. It describes sectors of healthcare practice, the relevance of medicinal plant use, medicalization and trends towards Western biomedicine, tensions between traditional and Western medicine, and factors affecting the loss or preservation of local medicinal plant knowledge. Overall, the lesson is approximately 1300 words. We embed a link[6] to a short (3-minute) video showing Guatemalan Maya women harvesting and discussing the uses of medicinal plants. After reading both lessons, students read a scholarly article about Maya peoples' use of medicinal plants and Western scientific support for those uses.

To conclude, we ask students, "What would help maintain medicinal plant knowledge and practice for the benefit of local people and global populations?" To address this question, students write a short (500-word) essay that weaves together cultural and ecological considerations into an imaginary development project that preserves Mayan ethnopharmacological knowledge (assignment instructions are in Appendix 4). Students are to propose a project that is culturally and ecologically appropriate. Once students write the essay, they post it in an online discussion board to share it with classmates.

Each student must then respond to two other students' essay posts. In their response, they are to acknowledge how the students' project responded to cultural and ecological considerations and offer constructive criticism for how the project could be improved. Finally, the class met in person to review the development projects that students suggested, discuss the pros and cons of each proposed project, and reflect on the process of completing the lesson.

Resources Needed

This lesson was available to students through Washington State University's Blackboard online platform. Any online platform would work for this lesson, but it should be one easily accessible to students (such as one already in use by the institution). The lesson takes approximately 3–4 hours to complete and, because this was not a strictly online course, students were given a day off from regular lecture schedule to complete it on their own time. The final, in-person discussion for this lesson happened during the regular course meeting time in the regular classroom. This final discussion required additional time beyond the 3–4 hours for the lesson material. If this were a solely online lesson, this final discussion format could be adjusted (more on this in a subsequent section).

Outcomes

Student Accomplishments

In this lesson, students demonstrated knowledge of the geography, culture, and ecology of Mesoamerica. They linked the importance of these foundational topics with each other, and with healthcare and the practice of medicine. Class members integrated knowledge and cultural sensitivity with their creative ideas to design a small development project aimed at preserving medicinal plant knowledge and practice in Maya communities. They accessed information, communicated their learning, and critically examined and peer-reviewed their own and classmates' ideas of culturally appropriate development proposals, in the online environment.

Developing a "Community of Inquiry"

In an online environment, like an in-person one, there is still a need for quality course content and a "community," or social interaction that guides, motivates, and assesses learning. For online courses to be highly effective learning systems, they require restructuring the typical, in-person, classroom teaching style (Means et al. 2009) rather than replicating it in a different environment (Brewer and Brewer 2015). Course content may stay relatively the same, yet instructors must pay extra attention to diminishing the "transactional distance" (Moore 2013), or the effort it takes to connect and communicate ideas, information, and a sense of community between people.

The "community of inquiry" model postulates that three intertwined branches of activity, or "presence," are necessary for successful online learning: cognitive, teaching, and social presence (Garrison 2009; Garrison et al. 1999). The "community of inquiry" framework emphasizes the communicative environment in which inquiry—that is, understanding, meaning-making, and collaborative engagement—take place (Garrison 2009). In this section, we will show how we used the "community of inquiry" framework to guide the instructional design of this lesson. In the next section, we reflect on this framework using student feedback and our perceptions of how the lesson implementation met the original objectives.

Cognitive Presence

"Cognitive presence" refers to the effect and quality of the content, methods of instruction and assessment, and overall learning that students achieve within a course. It refers to the process by which "learners are tasked with a problem or issue and, through iteration between discourse and reflection . . . construct meaning and confirm understanding" (Garrison 2009:353). This lesson achieves cognitive presence by presenting information in dynamic ways and assigning activities and assessments that are meaningful to the content and student learning (Brewer and Brewer 2015).

To present information dynamically, we include multi-media content in both lessons. Multi-media content in our lesson refers to photos and short videos that illustrate course content and link to websites with interactive and relevant material. For example, when discussing the migration of Mesoamericans to the US in the notes, there is a link to a Migration Policy Institute webpage that shows a graph of the US immigrant population by country of birth. Students may select Mesoamerican countries from the menu and results are then displayed immediately in an easy-to-read graph, helping students get a sense for the rates of immigration by country, selecting whichever countries interest them. Using mixed-media provides different styles of content presentation to engage students' different styles of learning and encourages students to internalize the messages of the material. Respecting and encouraging these aspects of active and diverse ways of learning are essential in education, regardless of the level or environment (Chickering and Gamson 1987).

Several aspects of this lesson encourage "deep learning"—an evolving concept referring to the cognitive depth and level of meaning derived from a learning experience (Czerkawski 2014). We scaffolded, or layered, the material, with the most basic and superficial first (culture, ecology, history, current issues) and the more complex material last (health care, ethnopharmacology). We similarly scaffolded the assessments, with a quiz that requires review of Lesson 1, and which also provides context for both Lesson 2 and the writing and discussion assignment. The writing and discussion assignment encourages interaction and evaluation among peers, which has been shown to support deeper, active learning (Sitthiworachart and Joy 2008) and higher levels of student satisfaction and success (Swan 2002).

Furthermore, the activities and assessments require students to demonstrate all parts of Bloom's (1956) classic learning taxonomy: knowledge, comprehension, application, analysis, synthesis, and evaluation. Students demonstrate knowledge and comprehension in the quiz. They apply their knowledge and comprehension to a real-world problem by suggesting a culturally-sensitive development project for traditional ethnopharmacological knowledge preservation. They analyze, synthesize, and evaluate by the process of peer review and group discussion. These educational objectives contribute to a sense of deep learning and cognitive presence within this lesson.

Social Presence

Social presence—the building of a shared community with a common purpose—is another critical aspect of successful online learning (Garrison 2009; Garrison et al. 1999; Whiteside 2015). Social presence includes a range of emotional attributes (sense of trust, cohesion, affect) that grows with instructor-to-student and student-to-student interaction (Whiteside 2015). Without the physical presence and interactive behaviors found within a classroom experience, course designers must develop social presence in the online environment (as Short et al. 1976 found with earlier technology). Pragmatically, this happens through email, discussion boards, video or audio lectures, and conference calls. Ways of relating that build social presence include acknowledging others, affective expressions (i.e., "I like your idea,") collaborative learning, and frequent, meaningful interaction (Hill et al. 2001; Swan 2002).

In this case, the social presence in the lesson, *Culture, Ecology, and Health in Mesoamerica*, began in the typical, in-person classroom environment, as the students and their course instructor (MQ) interacted with one another for several weeks before completing the online lesson. The strictly online form of social presence came from the peer review discussion assignment, in which students expressed their ideas and personality as they shared their written work and constructively critiqued one another's work. We rekindled in-person social presence when the class met to reflect on the lesson. In a strictly online course, the social presence would develop over time solely through online activities and interaction.

Instructor Presence

Teaching presence integrates and structures the cognitive presence and social presence in the learning environment (Garrison 2009). There are three principal functions of teaching presence: design, facilitation, and direct instruction (Anderson et al. 2001). The instructor must plan and present engaging content, encourage student interaction and collaboration, address student challenges and concerns, and provide feedback and assessment, all in a timely and consistent way (Garrison 2009; Goldman 2011). Here, MQ had already established instructor presence in-person throughout several weeks before transitioning into the online assignment. The online lesson nevertheless communicated instructor presence by providing clear expectations, an organized and intuitive format, and timely response to questions and concerns, all of which enhance student satisfaction regarding the instructor's role (Goldman 2011; Sheridan and Kelly 2010).

Additional Considerations

In addition to the cognitive, social, and instructor presences described within the community of inquiry framework above, there are other important considerations for the design and implementation of successful online lessons. More specifically, academic integrity and accessibility play large roles in the online environment.

Academic integrity is perhaps a more significant concern in online courses compared to in-person ones because of the potential lack of student visibility (Kitahara and Westfall 2007). On the other hand, students who tend not to speak up in classroom-based courses may feel more comfortable expressing themselves in the online, text-based environment (Brewer and Brewer 2015). Either way, academic integrity is an essential aim in any learning environment. We first addressed academic integrity by using a variety of assessment measures (e.g., quiz, essay, peer-review discussion) which requires students to demonstrate learning in different and original ways (WCET 2009). Second, students had to integrate various sections of the lesson material into their understanding and voice by completing an essay. This essay required students to write a brief review that explained the lesson content and to use their understanding of lesson material to design an original project for traditional ecological knowledge preservation that respects local customs, culture, and needs. This type of assessment requires "higher order thinking" (Olt 2009:223), or the use of critical analysis that blends information with reflection and application. Third, we required students to provide their classmates with peer review of their essays, which encourages originality and synthesis of lesson material (WCET 2009).

Finally, another important consideration in developing and implementing online courses is their accessibility or "creating a learning environment that works for all learners" (Perez 2015:60). Designing accessible online lesson content benefits all learners, not just those with disabilities (WebAIM 2016). Ways we improved accessibility in our lesson include using clear, consistent formatting, and a 12-point san-serif font in black ink on a white background display. Although simple, these considerations significantly increase the ability for learners to see, understand, and follow along with lesson material (WebAIM 2016).

Reflection: Student Response and Instructor Perception

After students completed the online lesson material, we met with them to reflect on the ideas they generated in their essays and to gain feedback on the lesson material and process of interacting online. We asked students to tell us the things they found most important or helpful in this lesson. We listed their ideas on the whiteboard and they discussed each. We did not quantify their degree of agreement; however, we report the items that generated verbal and visual agreement or interest. Student feedback and our perceptions of the success of this online lesson reflect the "community of inquiry" framework (Garrison 2009; Garrison et al. 1999). Now we turn our attention to specific examples of student response to our online lesson.

I. Cognitive Presence

Students expressed general appreciation for many cognitive aspects of our lesson:

- The "to-do" list, as it provided a concept map and clear expectations.
- Having separate documents for each section of the lesson (e.g., Lesson 1, Lesson 2, and scholarly article).
- Interactive tools (links, graphs, videos) added variety and relevance.
- Required reading of a scholarly article provided extra information and context and was appropriate for upper-level students.
- Essay required synthesis, creativity, and a recap of the information in the lessons.
- Quiz questions required review of lesson material and also helped the essay writing process.
- Short-answer quiz questions demanded synthesis of material.
- Assessments were relevant and purposeful, and did not feel like "busy work."
- The final discussion that reviewed and critiqued the development project proposals was an excellent way to close the lesson. If conducted solely online, students suggested that the instructor facilitate a final discussion in the discussion board. They could do this by posting a brief description of all of the development ideas proposed in the essays, and ask students to critique them according to their awareness of local cultural acceptability.

Some students critiqued the cognitive aspects of the lesson:

- The multiple-choice quiz questions were too easy.
- The quiz questions and the requirement for a brief cultural overview in the written assignment were too similar and possibly redundant. The redundancy could have been

improved by only requiring one assessment or by requiring that the essay *integrate* cultural overview into the development proposal, rather than introduce it separately.

- The lesson should be designed to fit the students' academic background better. For example, considering that they are Honors students, the lesson should be more rigorous regarding content and assessment.
- It would have been helpful to see more examples of anthropologists, their research, and development projects in which they participate.

We find several of the students' critiques worth addressing. We agree with them that the multiple-choice quiz questions were easy and likely did not demand a level of understanding or synthesis appropriate to their academic background as Honors students in an upper-level course. We *may* adjust the quiz to contain more short-answer questions, rather than multiple choice ones, to require more original and rigorous thought and reflection. Additionally, reducing the time limit on the quiz would encourage more command of the material. Still, the quiz serves as a checkpoint to ensure understanding of basic information before moving on to more complex lesson material, using the automatic grading function in our Blackboard software. The quiz is not the main means by which we assess learning. It is more important for the students to demonstrate their understanding of the material by integrating it into their essay. We consider addressing the issue of redundancy in assessments by adjusting the essay instructions to emphasize the integration, rather than the repetition, of basic cultural facts into students' development plans.

As instructors, we noticed that students were highly engaged in the lesson. They performed very well on the quiz, with 90% of students earning full credit (perhaps because, as we mentioned, the questions were not demanding enough, but also because Honors students are a select group of the university's high-achievers). Students wrote essays proposing the preservation of traditional ethnopharmacological knowledge that mostly met the requirements: they were well organized, used appropriate literature, sufficiently explained the cultural context and need for development, and suggested creative, culturally appropriate ideas for knowledge preservation. Some examples of these development proposals include creating books or guides that document knowledge in the local language; cultural resource centers that investigate, document, and share knowledge; online databases; video documentation; and incorporation of traditional medicine studies into Western biomedical training. Keeping the essay requirement short (~500 words) meant that 1) students had to focus on the main points and relevance of their suggestions, and 2) were able to read each other's work and provide peer review without a substantial investment in time and energy. Using the discussion board, students provided peer-review that was constructive while still addressing challenges and gaps. Of course, not all students excelled, but the vast majority of them met the requirements and our expectations, indicating the cognitive success of this lesson.

II. Social Presence

We noticed several indicators of strong social presence in the online lesson. The peer-review requirement in the discussion forum encouraged critical thinking and peer engagement. Students said that it was "helpful to see others' ideas." The flip side of this is that students may have copied each other's ideas rather than coming up with their own. To address this potential problem, in the future we could adjust the settings in the online course space to release all student essays for viewing at the same time, and after the submission deadline.

In their peer review, students constructively criticized, respectfully disagreed, kindly identified gaps and challenges, gave helpful suggestions, and were generally considerate and thoughtful in their responses. In a couple of cases, students overly praised each other's work without any truly meaningful feedback. Other times, students only criticized stylistic or grammatical issues. While these comments may also be educational for the writers, we realize that providing a very clear rubric for online discussion etiquette would be helpful to guide students away from superficial statements and toward more meaningful responses.

III. Teaching Presence

We found teaching presence best exemplified by setting clear expectations for student responsibilities and our facilitation of a review discussion at the end of the lesson. As mentioned before, we conducted the review discussion in-person, though we do not attribute its success as a learning tool to the in-person environment. Instead, we see this activity contributing to the cognitive and social presence of the lesson because it provided a review of the lessons in the material and created connections in the learning community. We could have facilitated review of the students' development proposals and critical discussion just as successfully in an online discussion board.

Conclusions and Future Prospects

Revisiting the Institution and Course

Our online lesson, *Culture, Ecology, and Health in Mesoamerica*, fit well with topics in the 300-level Honors College course, *Human Health and Healing across Cultures*. The lesson was scheduled in the week that dealt with ethnopharmacology, and as such, provided students with an applied understanding of the field as it relates to the Mesoamerican cultural and ecological context. Students completed readings and watched a film relating to ethnopharmacology before our online lesson (we list these resources in the Notes section below).

Human Health and Healing across Cultures is a course conducted in a typical classroom environment. For this course, students make use of the online Blackboard platform to access assigned readings and submit weekly assignments. Our lesson was the only portion of the course conducted entirely online. Students responded well to the online lesson, probably in part because of their familiarity with the online learning platform, Blackboard, and also in part because the lesson topics were a natural extension of other course material.

Adapting the Lesson to Other Courses or Institutions

We believe that this lesson could be readily adapted to other courses and institutions, so long as the objectives and topics of the course align reasonably well. The lesson could fit with courses discussing Latin America, developing countries, health and healing, ethnobiology, human ecology, international development, or cultural studies. To facilitate the adaptation of this course to other contexts and provide general tips for online course planning, see the figure below. For our full lesson plan, see Appendices 4–6.

Another consideration for adapting this lesson to other courses or institutions is students' level of familiarity with the online learning environment in general, and the institution's online platform (i.e., Blackboard, Angel) in particular. Experienced online learners will likely find this lesson format more straightforward to understand. Lack of familiarity is not an obstacle to online learning, however, so long as the preparation builds a compelling learning experience that fosters cognitive, social, and teaching presence.

In Table 8.2 below, we outline our suggestions for developing online lessons. The methods we mention first in each section apply to online courses generally. Those that follow apply to teaching online ethnobiology and human ecology courses, more specifically. For ethnobiology and human ecology online course planning, methods should reflect the particular topic of the lesson at hand. For example, in a lesson on ethno-ornithology, audio or video clips of birds (within the context of lecture notes or other reading) would be a relevant content item. Furthermore, requiring students to observe birds or human-bird interactions in the real-world could be a useful assignment that is then followed up on within a discussion in the online class.

Final Thoughts on Online Learning

Online learning is on the rise, and has many potential benefits and challenges. Benefits of online learning are that students may achieve academic objectives with greater scheduling flexibility, and students can focus on the message of the content and interaction, rather than who is saying it (Ocker and Yaverbaum 1999). Students have more flexibility in the pacing of the course, even if there is a required schedule (Brewer and Brewer 2015). Students may review lesson materials as frequently as they choose, given the parameters of the schedule, as opposed to classroom-based courses that present a lesson or lecture once and then move on. Hence, online/distance learning has the potential to increase student diversity, increasing equity by widening access to students who may be employed, geographically remote, or otherwise limited by disability or family responsibilities (Stone et al. 2016).

Table 8.2. General suggestions for online lessons/courses.

Aim	Methods to Support the Aim	Relevant Literature
Build cognitive presence	• Explain course aims and methods to achieve them • Provide clear, engaging content • Use mixed-media (photos, videos, graphs, figures, links to related material) For Ethnobiology and Hunam Ecology: • Adapt the same content as in-person classes with extra attention to multi-media to illustrate points • Include assignments that engage students in the "real" world (i.e., observations, interviews, outdoor activities)	Brewer and Brewer 2015 Bloom 1956 Chickering and Gamson 1987 Czerkawski 2014 Garrison 2009 Sitthiworachart and Joy 2008
Build social presence	• Communicate regularly and often • Provide an ungraded discussion forum for student communication • Require interaction (through the discussion board or conference calls) For Ethnobiology and Hunam Ecology: • Include assignments that require interaction (i.e., interviews with community members or experts, participant observation, photography or video assignments) • Highlight an expert's work by showing a video or conducting synchronous videoconferencing	Garrison 2009 Hill et al. 2001 Moore 2013 Short et al. 1976 Swan 2002 Whiteside 2015
Build teaching presence	• Communicate clear expectations • Provide timely response to student needs and concerns • Provide timely feedback For Ethnobiology and Hunam Ecology: • Showcase instructor's own work as examples of relevant content	Anderson et al. 2001 Garrison 2009 Goldman 2011 Swan 2002 Sheridan and Kelly 2010
Enhance accessibility	• Use a consistent formatting style throughout • Display text in black ink on a white or light background • Use a san-serif font (Arial, Helvetica, Geneva) to reduce visual fatigue • If using video or audio lectures, also provide a written transcript	Perez 2015 WebAIM 2016
Promote academic integrity	• Require scaffolded assignments (draft or bibliography, followed by final draft) • Require peer review • Use a test bank and randomized test questions • Set a reasonable time limit for online exams	Kitahara and Westfall 2007 Olt 2009 WCET 2009

Challenges to online learning center around the "community of inquiry" framework (Garrison 2009; Garrison et al. 1999). When cognitive, social, and instructor presence are engaged, students excel. When one or more aspect is missing, student success and satisfaction wane (Goldman 2011; Sheridan and Kelly 2010; Swan 2002; Whiteside 2015). We foresee a unique challenge to adapting ethnobiology and human ecology courses to the online environment in the sense that the content of such courses lends itself well to hands-on, experiential, and "real-world" learning. In transferring such course content to the online environment, care should be taken to preserve the strengths of these disciplines. With dynamic content, an interactive community, and appropriate design and facilitation, ethnobiology and human ecology learning will be able to flourish in the online environment.

Notes

Links within Our Online Lesson:*

*all links are current at the time of this manuscript's writing.

1. Link 1: From Milpa to Mesa (The Perennial Plate 2016). URL: https://www.theperennialplate.com/episodes/2017/02/episode-167-from-milpa-to-mesa/.
2. Link 2: Surveying the Pre-Columbian (Leibsohn and Mundy 2005). URL: https://www.smith.edu/vistas/vistas_web/units/surv_precol.htm.
3. Link 3: History of Central America and the Caribbean—Animated Map (Esemono 2012) URL: https://www.youtube.com/watch?v=bUoypQ0ykwI.
4. Link 4: Countries of Birth for U.S. Immigrants, 1960–Present (Migration Policy Institute 2018). URL: https://www.migrationpolicy.org/programs/data-hub/charts/immigrants-countries-birth-over-time?width=1000&height=850&iframe=true
5. Link 5: Drugs in Mexico (The Economist 2019). URL: https://www.economist.com/node/21555750/mobile-frameless.
6. Link 6: Medicinal Plant Tour in San Juan La Laguna (Rehder 2009). URL: https://www.youtube.com/watch?v=lvtGE7MwIpU.

Related Resources for an Ethnopharmacology Lesson:

Prior reading: Balancing the System: Humoral Medicine and Food in the Commonwealth of Dominica (Quinlan and Quinlan 2006).

Prior film: The Shaman's Apprentice (with Mark Plotkin; Smith 2001)

Reading for online lesson: Medical Potential of Plants Used by the Q'eqchi Maya of Livingston, Guatemala for the Treatment of Women's Health Complaints (Michel et al. 2007).

References Cited

Anderson, T., L. Rourke, D.R. Garrison, and W. Archer. 2001. Assessing Teaching Presence in a Computer Conferencing Context. *Journal of Asynchronous Learning Networks* 5(2). [online] DOI=10.1.1.409.9114. URL: http://citeseerx.ist.psu.edu/viewdoc/download?doi=10.1.1.409.9114&rep=rep1&type=pdf. Accessed on July 5, 2019.

Bloom, B.S., ed. 1956. *Taxonomy of Educational Objectives. Vol. 1: Cognitive Domain.* McKay, New York.

Brewer, P.E., and E.C. Brewer. 2015. Pedagogical Perspectives for the Online Education Skeptic. *Journal on Excellence in College Teaching* 26(1):29–52.

Chickering, A.W., and Z.F. Gamson. 1987. Seven Principles for Good Practice in Undergraduate Education. *AAHE (American Association of Higher Education) Bulletin* 3:1–6.

Czerkawski, B.C. 2014. Designing Deeper Learning Experiences for Online Instruction. *Journal of Interactive Online Learning* 13(2):29–40.

Economist, The. 2019. Drugs in Mexico [web page]. URL: https://www.economist.com/node/21555750/mobile-frameless. Accessed July 2019.

Esemono. 2012. History of Central America and the Caribbean—Animated Map: Political Evolution of Central America and the Caribbean 1700–1983. EarthDirect (Public Domain). URL: https://www.youtube.com/watch?v=bUoypQ0ykwI. Accessed on July 4, 2019.

Garrison, D.R. 2009. Communities of Inquiry in Online Learning. In *Encyclopedia of Distance Learning, Second Edition*, edited by P. Rogers, G. Berg, J. Boettcher, C. Howard, L. Justice, and K. Schenk, pp. 352–355. Information Science Reference (IGI Global), Hershey, PA.

Garrison, D.R., T. Anderson, and W. Archer. 1999. Critical Inquiry in a Text-based Environment: Computer Conferencing in Higher Education. *The Internet and Higher Education* 2(2–3):87–105.

Goldman, Z. 2011. Balancing Quality and Workload in Asynchronous Online Discussions: A Win-Win Approach for Students and Instructors. *Journal of Online Learning and Teaching* 7(2):313–323.

Hill, J.R., S. Han, and A. Raven. 2001. Build It and They Will Stay: A Research-Based Model for Creating Community in Web-Based Learning Environments. In *2001 Annual Proceedings—Atlanta of Selected Research and Development Papers Presented at The National Convention of the Association for Educational Communications and Technology*, edited by M. Crawford and M. Simonson, pp. 192–199. Association for Educational Communications and Technology, Bloomington, IN.

Kitahara, R.T., and F. Westfall. 2007. Promoting Academic Integrity in Online Distance Learning Courses. *MERLOT Journal of Online Learning and Teaching* 3(3):265–276.

Leibsohn, D., and B. Mundy. 2005. Surveying the Pre-Columbian. Vistas, Spanish American Visual Culture, 1520–1820 [web page]. URL: https://www.smith.edu/vistas/vistas_web/units/surv_precol.htm. Accessed on July 5, 2019.

Means, B., Y. Toyama, R. Murphy, M. Bakia, and K. Jones. 2009. *Evaluation of Evidence-based Practices in Online Learning: A Meta-analysis and Review of Online Learning Studies.* U.S. Department of Education, Jessup, MD.

Michel, J., R.E. Duarte, J.L. Bolton, Y. Huang, A. Caceres, M. Veliz, and G.B. Mahady. 2007. Medical Potential of Plants Used by the Q'eqchi Maya of Livingston, Guatemala for the Treatment of Women's Health Complaints. *Journal of Ethnopharmacology* 114(1):92–101.

Migration Policy Institute (MPI). 2018. Countries of Birth for U.S. Immigrants, 1960-Present. Migration Policy Institute [web page]. URL: https://www.migrationpolicy.org/programs/data-hub/charts/immigrants-countries-birth-over-time?width=1000&height=850&iframe=true. Accessed on July 5, 2019.

Moore, M.G. 2013. The Theory of Transactional Distance. In *Handbook of Distance Education*, edited by M. Moore, pp. 84–103. Routledge, Philadelphia.

Ocker, R.J., and G.J. Yaverbaum. 1999. Asynchronous Computer-Mediated Communication Versus Face-to-Face Collaboration: Results on Student Learning, Quality and Satisfaction. *Group Decision and Negotiation* 8(5):427–440.

Olt, M.R. 2009. Seven Strategies for Plagiarism-Proofing Discussion Threads in Online Courses. *MERLOT Journal of Online Learning and Teaching* 5(2):222–230.

Perez, L. 2015. From Accommodation to Accessibility: Creating Learning Environments that Work for All. *EDUCAUSE Review* 50(3):60–61.

Perennial Plate. 2016. From Milpa to Mesa [Film]. Perennial Plate Documentary Series, Season 4, Episode 167. Masienda and the Perennial Plate. URL: https://www.theperennialplate.com/episodes/2017/02/episode-167-from-milpa-to-mesa/. Accessed July 5, 2019.

Quinlan, M.B., and R.J. Quinlan. 2006. Balancing the System: Humoral Medicine and Food in the Commonwealth of Dominica. In *Eating and Healing: Traditional Food as Medicine,* edited by A. Pieroni and L.L. Price, pp. 197–212. Haworth Press, Binghamton, NY.

Rehder, S. 2009. Medicinal Plant Tour in San Juan La Laguna [Film]. Fundacion Salar, Guatemala. URL: https://www.youtube.com/watch?v=lvtGE7MwIpU. Accessed July 2019.

Sheridan, K., and M.A. Kelly. 2010. The Indicators of Instructor Presence that Are Important to Students in Online Courses. *Journal of Online Learning and Teaching* 6(4):767–779.

Short, J., E. William, and B. Christie. 1976. *The Social Psychology of Telecommunications.* Wiley, Toronto, ON.

Sitthiworachart, J., and M. Joy. 2008. Computer Support of Effective Peer Assessment in An Undergraduate Programming Class. *Journal of Computer Assisted Learning* 24(3):217–231.

Smith, M., dir. 2001. The Shaman's Apprentice [Film]. Miranda Productions, Boulder, CO.

Stone, C., S. O'Shea, J. May, J. Delahunty, and Z. Partington. 2016. Opportunity through Online Learning: Experiences of First-in-Family Students in Online Open-Entry Higher Education. *Australian Journal of Adult Learning* 56(2):146–169.

Swan, K. 2002. Building Learning Communities in Online Courses: The Importance of Interaction. *Education, Communication & Information* 2(1):23–49.

WebAIM. 2016. Introduction to Web Accessibility [web page]. *Web Accessibility in Mind.* URL: https://webaim.org/intro/. Accessed on October 17, 2018.

WCET. 2009. Best Practice Strategies to Promote Academic Integrity in Online Education, Version 2.0 [web page]. Western Interstate Commission for Higher Education Cooperative for Educational Technologies. URL: https://wcet.wiche.edu/focus-areas/student-success/academic-integrity. Accessed on October 17, 2018.

Whiteside, A.L. 2015. Introducing the Social Presence Model to Explore Online and Blended Learning Experiences. *Online Learning* 19(2):n2.

Chapter 9.

Training Environmental Professionals through a Collaborative Experiential Learning Capstone

Daniela J. Shebitz[1] and Michael Edelstein[2]

Chapter Summary

The senior capstone is one of the key high impact practices to student learning (Kuh 2008) and should complete the foundation of knowledge that students construct throughout their undergraduate education. In this chapter, we present a structure for an experiential learning capstone course with a primary objective of encouraging students to develop their ability to address and analyze complex problems and to creatively apply their acquired knowledge and skills to real-life situations. An additional objective is to increase student connections to the greater community. Through collaborative assignments and projects, students work as a consulting firm to design and implement an integrative project that addresses a need for a particular "client" and builds upon the knowledge they've gained throughout their time as an undergraduate student. The course structure evolution at Ramapo College and its adoption at Kean University is discussed through the professors' perspectives. One project that was the focus of a capstone at Ramapo College is presented as a case study that links the students to a local Indigenous culture, The Ramapough-Lenape Nation's Ringwood Turtle Clan. We show that capstone projects bridging science, policy, and the public not only positively benefit students by preparing them for the real world, but additionally benefit the greater community. This approach to the professional empowerment of upper-level undergraduates working with community partners is applicable to multiple fields and projects

1 Kean University, School of Environmental and Sustainability Sciences
2 Ramapo College of New Jersey, Programs in Environmental and Sustainability Studies

Introduction

All too often, students graduate college without an understanding of how to translate their course work into professional experience. The absence of a clear connection between their courses to each other and to their professional goals not only affects their résumé and ability to get jobs, but also their level of confidence as recent alumni. While students may be graduating with high GPAs, the National Association of Colleges and Employers (NACE; 2013:4) states that successful employment upon graduation is often dependent on general, transferable professional skills including: 1) The ability to work in a team setting; 2) proficiency in problem solving and decision making, 3) good organizational skills, and 4) the ability to communicate skillfully. Increasingly, graduates are being called upon to work in cross functional teams, and the skills learned in integrative capstone courses are key for their preparation (Franchetti et al. 2012). This need for collaboration is especially strong in the area of human ecology where only through a transdisciplinary effort can we start to understand the complex interrelationships between individuals, communities, and their environments.

There is concern that traditional university programs lack the teaching methods to turn students into innovative and creative leaders who can integrate skills and fields of study (Franchetti and Ariss 2016). Moreover, it is uncommon to find information presented in the context of multidisciplinary or socio-economic perspectives. The failure of instruction to move beyond traditional science to interdisciplinary approaches leaves majors ill-prepared to perform well early in their career (Brewer and Smith 2011:2–6; Committee on Undergraduate Biology Education 2003:10–14; Shebitz et al. 2017a).

The Association of American Colleges and Universities recognizes a senior capstone as one of the key high impact practices to student learning (Kuh 2008). Just as an architectural capstone is the highest stone on a building that completes the structure, the senior capstone course should complete the building that students construct throughout their years of education (Appleby et al. 2016). These culminating experiences require students nearing the end of their college years to create a project that integrates and applies what they've learned, such capstones vary widely in terms of structure, content, and assignments (Grahe and Hauhart 2013). While larger schools and those offering graduate programs are less likely to offer capstone courses (Hauhart and Grahe 2012), programs in Environmental Science, Environmental Studies, and Sustainability among other fields of study are calling for interdisciplinary teamwork to develop solutions as part of their capstone course (Latif et al. 2015; Shebitz et al. 2017a).

In this chapter, we present a structure for an experiential learning capstone course that was developed at Ramapo College of New Jersey and adopted by Kean University, both of which are public, comprehensive higher education institutions located in the New York Metropolitan Area with a primary focus on the teaching and learning. The primary objective of the capstone structure presented in this chapter is to encourage students to develop their ability to address and analyze complex problems and to creatively apply their acquired knowledge and skills to real-life situations. An additional objective is to increase student connections to the greater community. Through collaborative assignments and projects, students work as a consulting firm to design and implement an integrative project that addresses a need for a particular "client" and builds upon the knowledge they've gained throughout their time as an undergraduate student. One project that was the focus of a capstone at Ramapo College is presented as a case study in this *Educational Contributions in Ethnobiology* edited volume for two reasons: 1) it is a terrific example of how university students can work with local Indigenous communities, and 2) it was the project that engaged both authors of this chapter and laid a foundation for a similar framework at Kean University.

The Course Evolution at Ramapo College

From Mike Edelstein: It may be useful to begin with my own background and how I became oriented toward creating a practice course in environmental study. Shortly after beginning my Ph.D. work in Social Psychology in 1970, I adopted the new field of Environmental Psychology. Unlike most psychologists, I was a contextualist, interdisciplinary in focus and interested in cutting-edge, field-oriented action research aimed at changing the world. After offering my first Environmental Psychology course in SUNY Buffalo's Psychology Department, I was asked to join their School of Architecture and Environmental Design. For the next three years, while completing my Ph.D., I worked in a collaborative program team-teaching field-oriented studio courses, a stark shift from the traditional courses I had myself taken during my education. The focus on participatory teaching and design meshed with my Ph.D. work in group dynamics and organizations, culminating in completion of the National Training Lab's two-year Graduate School Professional Development Program. As a result, I was adept at creating dynamic classrooms reflecting my process-perspective of the world.

In 1974, I joined the new School of Environmental Studies at Ramapo College of New Jersey, an institution whose unique mission was to offer such interdisciplinary programs through experientially-oriented multi-cultural and global curricula. The field of Environmental Studies was a reflection of the Environmental Era and specifically the piece of legislation that defined it, the National Environmental Policy Act (NEPA), signed on January 1, 1970 by President Nixon. NEPA had as its stated national mission to create harmony between humans and the environment, in part, through new efforts at interdisciplinary learning.

Ramapo's collaborative and cohesive faculty were drawn to this mission. Ramapo's School of Environmental Studies quickly became one of the foremost centers for undergraduate study of renewable energy, sustainable food, waste cycling, social ecology, environmental policy, and environmental psychology. Our student-built wind-powered Alternative Energy Center (AEC) featured a passive solar greenhouse (circa 1975), a solar schoolhouse, huge gardens, a tree farm and an orchard, and an entrepreneurial recycling center that developed lucrative markets for recovered materials that financed the whole center. Perhaps most importantly, the AEC created an active community of students who worked on projects and celebrated the environment. It also served to disseminate Ecological Literacy. Over time, most graduates of Ramapo and many community members and school children learned about sustainability through the center.

By the mid-1980s, under the dark shadow of the Reagan administration, the School of Environmental Studies was closed and a redesigned all-college Environmental Studies major was launched. With my colleagues, I broadened my focus to diffusing ecological literacy beyond the Environmental Studies major (Edelstein 2004a, 2009). I was also deeply involved in the emerging field of Social and Environmental Impact Assessment, created by NEPA's "action forcing mechanism," the Environmental Impact Assessment (EIA), specializing in psycho-social impact assessment (Edelstein 2004b). It became clear to me that, if there is a methodology—or meta-methodology—for the environmental field, it is EIA. EIA creates an interdisciplinary space where theory, practice and policy intersect. For the reshaped major, nothing could be more appropriate as a capstone than a course based on NEPA and EIA. I collaborated with a colleague to create this capstone, a course called Environmental Assessment, but then took it over. Eventually a second course, Field Project in Environmental Assessment, was added as a co-requisite. Although they blend together, one course mixes lecture on EIA with numerous student projects and presentations while the second is completely collaborative and project-oriented. A student consulting firm is formed, managed by a team of student leaders. Students collaborate to complete a real EIA project during the semester, responding to an RFP written with real clients who participate and support the class effort. Both courses fulfill the college's writing-intensive requirement. Class is either run for an entire day each week or for two full mornings. Although designed for 15 students, it has run with as many as 34, depending upon how many students plan to graduate that year. I have now taught this capstone for some 35 years. In contrast to conventional academic classes, the real-time, real-world nature of Environmental Assessment and the need for students to bond, organize, and collaborate on an open-ended research project makes this fluid learning experience a wild ride for me as well as the students.

Adoption Aspects of the Capstone Structure at Kean University:

From Daniela Shebitz: With a background in linking ethnobiology and restoration ecology, I started my career as a faculty member of Kean University's Biology Department in 2006. Kean University has been recognized by Diversity Inc. as being one of the top colleges and universities in terms of diverse student populations. The University has a high percentage of students that are the first in their families to attend college and the vast majority of students are awarded need-based financial aid in the form of loans, scholarships, and grants. Within this framework, part of my teaching load through their biology program is the three-credit senior capstone. The format of course allows the faculty to select a theme for the semester and for the students to write a literature review paper and give an oral presentation relating to the theme. Early on, I recognized that while this individualized learning strengthened writing and oral communication skills, it did not address the need for capstones to strengthen students' ability to work in a team setting and develop proficiency in problem solving and decision making. In addition, this single-discipline approach is usually not effective for students from underrepresented groups, whose cultures may emphasize learning through direct engagement, a multidisciplinary perspective, and active learning (Shebitz et al. 2017a).

Within the past six years, we started a new School of Environmental and Sustainability Sciences, with its own three-credit capstone course for Bachelor of Science Majors entitled "Independent Practicum for Sustainability Sciences." Although the emphasis of this course was creating experiential learning opportunities to assist in a transition from college student to professional, the learning was mostly done individually. Then, in 2014, I was invited by Mike Edelstein to join a capstone meeting and presentations for the project that is discussed in the case study of this chapter. Seeing how the Ramapo upperclassmen evolved into professionals who collaborated with the community left a lasting impression on me.

Since then, Kean's capstone for the Environmental Biology, Environmental Science, and Sustainability Science majors has incorporated the Ramapo model of teams working as an environmental consulting firm with a client. However, Kean does not always require an Environmental Impact Statement (EIS: a product of an EIA), and, in contrast to Ramapo's eight-credit package, at Kean, the class is restricted to one three-credit time period, requiring the vast majority of work to be conducted outside of class time. With an emphasis on professional development, the students use this course as a transition from college to their careers in the environmental field, not only through the project, but also through lessons on searching for the perfect job or graduate school, resume building, cover letter writing, and interview skills.

Description of the Learning Activity

According to the course plan Edelstein developed for Ramapo and now utilized by Kean, students create an environmental consulting firm, working for real clients on a real project. As a result, the faculty member must negotiate with clients and develop a Request for Proposals (RFP) before the term begins. Clients are responsible for assisting the students with information and guidance, attending a scoping hearing where their questions are identified, the hearing on the draft project, and the end of semester presentation of the final project. Frequently, the clients also help arrange and guide field tours, make connections for field research, and arrange for access to affected stakeholders for interviews. Their role is critical. At the end of the course, the clients become owners of the final project.

The formation of the student consulting firm is the foundation of the project. Students post their résumés online (through Moodle, Blackboard, or a Google Site) prior to the opening class. They begin taking time to get to know each other, including the experience and skills each person brings to the table. As a first act, students are given the task to come up with a name for their new firm. Based on this interaction, the group selects officers for the firm: one or more Project Managers (PM), one or more editors, a graphics/GIS/web team, and a librarian. Under the direction of the new PM(s), the class next divides into specialized teams. For example, when conducting an EIA, there are Physical, Ecological, and Social Impact teams, with each individual matched to an indicator of impact pertinent to their team's focus. Team managers and editors are next selected.

Once the firm is organized, the clients are invited to present their RFP to the class. The firm's first task is to prepare a proposal in response to the RFP. The proposal must show an understanding of what the RFP demands and the clients want, demonstrate the firm's competence to complete the project, and the ability of the firm to undertake the project by presenting the qualifications and organization of the firm. Such proposals serve as a dry run for collaboration within the firm and for officers to carry out their roles, help to build confidence in the ability of the firm to tackle the huge project ahead, and forces the class members to be very clear about the meaning of the RFP. A logo for the firm is also created to adorn the proposal.

Subsequently, a formal public hearing is organized by the instructor to clarify the scope of the project. Students present key questions to be addressed by their indicator; the list of questions is then embellished and critiqued by clients, fellow students, and community members. The instructor then prepares a final scoping document that is shared with the clients and class. It becomes the working agenda for the project and semester.

The project format requires a rather unconventional role for students and faculty. Once the firm is launched, the class is guided by the goals of the project and run by team managers and editors. The instructor coordinates with clients, manages and runs the various hearings, guides and supports the firm management, and intervenes when necessary. Individual students collaborate with their teammates on their projects. Class time involves meeting collectively as a firm or as teams, "putting out fires," and offering feedback and guidance. The instructor gives iterative feedback on all submissions and on the evolving EIA. The art of facilitating the class requires that the instructor rely on student initiative, offering detailed guidance as needed, while not getting drawn into doing the project for the students.

For the lecture portion of the capstone at Ramapo, a textbook by Marriott (1997) was utilized that closely fits the course format. For the project courses on both campuses, Levi's (2010) guide to working in collaborative task groups was used. Periodic group and organizational assessments are used to identify problems demanding correction. Most of these problems are relatively easy to fix. For example, communication may not be flowing well enough between project managers and the teams. Or a group communication site may be hard for some to access. Students do periodic diagnostic assessments of the firm's process which are then discussed in class where the issues are resolved.

Diverse projects have been used in the Ramapo capstone, including these:

1. A number of studies anticipated and then measured the impacts of expanding I-287 from Mahwah to Boonton, NJ, crossing the Ramapo campus, forcing the erection of

a sound barrier to protect the college and stopping a truck weigh station proposed for the Ramapo campus.

2. The cumulative effects of gas pipelines on the New Jersey Highlands was assessed; several studies have been made of individual and cumulative development projects in the Ramapo River watershed.
3. Potential impacts of the New Jersey Highlands Act and creation of the Highlands Commission were assessed.
4. Among several studies of the Ramapo campus, a sustainability assessment and greenhouse gas inventory were undertaken, impacts of master plan revisions were done, and an assessment of the college's emergency response plan was made.
5. Studies of the impact of environmental contamination on affected communities included work in Pompton Lakes and Ringwood, New Jersey as well as Lower Manhattan after the World Trade Center Disaster.

Among Kean's projects have been:

1. The economic, ecological, and cultural impacts of removing three dams along a river to restore native fish populations and ecological function was assessed for Trout Unlimted, Inc.
2. Plans were developed for a pollinator garden and bat habitat to be funded and installed at the Staten Island Zoo.
3. A stream daylighting plan and EIS based on GIS, on-the-ground surveys, and NJ Department of Environmental Protection guidelines was created.
4. A trail system, educational signage, and environmental education curricula for an urban greenspace near campus was created for Groundwork Elizabeth in an effort to get community members outside and to appreciate the importance of an underutilized urban park.

In the next section, we highlight an example of the capstone project from Ramapo College that was introduced earlier.

Example of a Human Ecology Capstone Linking the Students to a Local Indigenous Culture

In this section, we present a case study involving Ramapo College seniors in the Environmental Assessment capstone working as a firm to address an RFP prepared by the instructor in consultation with a local Indigenous community of northern New Jersey, The Turtle Clan of the Ramapough-Lenape Tribe. Their Upper Ringwood community was the scene of extensive dumping of hazardous paint sludge by Ford Motor Company in the 1950s and 1960s and a failed Superfund cleanup that eventually required that the site be relisted to start over again. Two abandoned iron mines in Upper Ringwood were filled with waste from Ford, and Ford wastes were dumped in a nearby landfill. Additional paint sludge and other wastes were deposited throughout the town and region. Turtle Clan members and other residents continue to live among and atop of these sites. The second Superfund process initially targeted "cleanup" of the mines and landfill, although much of the waste was to remain in situ. The question facing the student firm was whether Upper Ringwood is habitable and what the impacts of relocation of the community might be.

At the time the RFP was written, the EPA was proceeding to identify remedial actions despite having produced an inadequate Environmental Justice (EJ) Assessment. Therefore, the class was directed to prepare a comprehensive EJ analysis considering the full impact of the proposed cleanup on the Native Americans living within the Superfund site. The three-page RFP given to the class during its first week of meeting in January 2014 read in part:

> The Ramapough-Lenape Nation's Ringwood Turtle Clan has issued an RFP for a study of Environmental Justice that will consider evidence at hand in a more cumulative and comprehensive manner than does the EPA EJ study. The 2014 Environmental Assessment class is invited to prepare . . . a full Environmental Justice Impact Assessment. It will assess all NEPA impacts of the Ford contamination with a particular interest in the secondary impacts related to Environmental Justice. In doing so, it will consider as risks all indicators of Physical, Ecological and . . . Social . . . Impact. Cumulative impacts and interactive effects will be assessed, including those of a short and long term nature, those entailing irreversible and irretrievable effects and those involving such areas of emerging impact as climate change.

The Project, accordingly, was intended to assess both the Ford contamination and also the impacts of the Superfund process.

The class named their consulting firm RISE (Research Investigation for Society and the Environment) and developed a logo (Figure 9.1). Within two weeks of receiving the RFP, a draft proposal had been iteratively improved in response to instructor feedback allowing RISE to submit a final 31-page proposal entitled "Superfund Remediation in Ringwood, New Jersey: A Proposal for Conducting an Environmental Impact Assessment of the Environmental Justice Impacts of Ford Motors and EPA Remediation of Ringwood Mines." It consisted of an overview of the proposed project and how it addressed the RFP. It also provided a detailed analysis of RISE's organizational structure (shown generically in Figure 9.2) and the

Figure 9.1. Logo for the 2014 Environmental Assessment class firm, Ramapo Colllege.

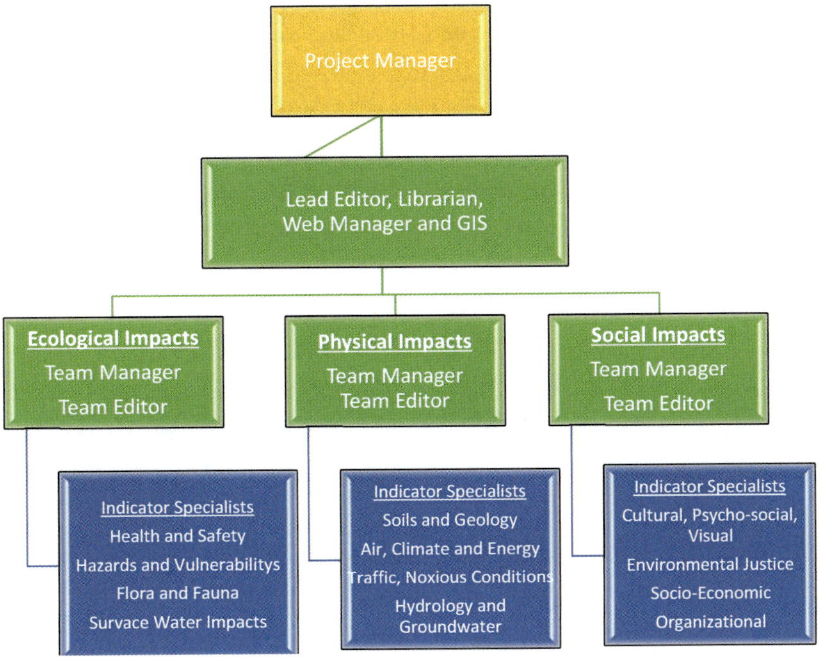

Figure 9.2. Organizational flow chart for RISE.

firm's capacity to undertake the project, appending brief Vitae for each student. The Proposal was subsequently accepted by the client.

To prepare for scoping the project, a field trip was made to Ringwood, New Jersey guided by Turtle Clan Chief Vincent Mann and other clan leaders. The class visited the three operable units in the Superfund cleanup (Cannon and Peters mine sites and the O'Conner Landfill), as well as residential properties affected by sink holes and gas mitigation. We hiked up to a tribal elder's home, an original iron mine worker residence, where the students spoke to a former iron worker and the elder and learned about impacts on community members' lifestyle and longevity. Such field trips make the issues under study vivid and create the relationships necessary for developing follow up fieldwork.

The three-hour long scoping hearing that took place two weeks later was attended by the clients. The Project Managers introduced the RFP, proposal, and firm. Each team leader introduced their staff members, who then presented a draft scope for their own individual impact indicator before it was further expanded in discussion with clients and classmates. After the hearing, the instructor prepared an official scoping document covering 24 indicators of impact, which subsequently served as a checklist for what was required in the final impact statement.

During March, field work was conducted on many indicators. IRB approval was received to interview Turtle Clan members, other residents and key informants by phone, mail, or email. When the Social Impact team had an extensive meeting with community members organized by Chief Mann, it became clear that elders were so tied to the land that they preferred to remain behind in the contaminated landscape. In contrast, many adult Turtle Clan members with school-aged children had already relocated on their own to protect their families from the contamination. Young adults who took part in the interviews described in detail how stigma against the Ramapoughs had affected their lives, particularly at school. Such stigma was clearly another type of contamination the community was dealing with.

The project report went through several submissions and revisions before a 163-page Draft EIS document was issued on April 28 and served as the basis for a hearing for clients and the public. Based upon feedback from the hearing, a Final EIS was issued ten days later. In their last class meeting, the class prepared a Findings Statement, prioritizing findings according to severity and significance. Around the same time, class members made presentations at the annual Ramapo River Conference and for a student research session. RISE made its final formal three-hour presentation to the clients, college community, and public on May 8.

One of the major problems with this course process is that the final document inevitably requires post-graduation editing. In this instance, the instructor helped RISE's chief editor get a law firm internship that summer where she worked on summarizing documents related to the Ringwood contamination for possible litigation. She volunteered to work with the instructor to do further edits on the FEIS before it was handed off to the clients. That editorial task was undertaken during the summer of 2014 and completed in early October. The Final EIS was 195 pages long plus a 28-page appendix, 12 tables, and 47 figures. It concluded, overall, that the EPA had indeed failed to examine disproportionate environmental impacts in its EJ analysis. EPA had concluded that a lack of definitive health data made it impossible to determine if disproportionate risk was born by the Ramapough community, but then proceeded to do the Remedial Investigation Feasibility Study without collecting the missing health data. By employing a full EIS that examined secondary psycho-social and physical health effects related to all significant adverse impacts of the contamination and cleanup, the Ramapo College project could define disproportionate risk against all of the 24 different impact indicators studied, not just health in isolation. Thus, the conclusion:

> Using our much more comprehensive approach, RISE found a very different conclusion, namely that the Ramapoughs have been adversely and disproportionately impacted by Ford's dumping.

This alternative approach to identifying EJ impacts raised questions missed by EPA's review. First, a last minute switch of plans after the conclusion of the review process scrapped a plan to remove contaminated soils from the O'Connor landfill and restore a natural area where the Ramapoughs could hunt and gather safely. In its place, EPA accepted a proposal from Ford and the Town of Ringwood, also a Potentially Responsible Party (PRP), to move its recycling center to the site. The switch allowed contaminants to remain in situ under a covering slab upon which the Recycling Center would be placed, saving the PRPs many millions of dollars.

Not only did residents lose the use of a restored natural area, but now a new heavily trafficked waste-related activity was inserted into the heart of their residential area instead. Had the RISE EJ analysis been utilized by EPA, it would have created leverage to stay with the original remedial plan. Additionally, the EJ study led to a consideration of habitability never broached by EPA. RISE recognized that the organized and subsidized relocation of the Ramapoughs to safe tribal lands, even if opposed by some residents, might constitute the only way to assure elimination of health impacts that have decimated the Upper Ringwood population for decades

Given the significance of the RISE study, the instructor prepared his own summary of the findings and recommendations in July 2014 in an effort to influence EPA decision-making. Subsequently, he offered testimony on the EJ impacts before the EPA, giving full credit to the substantial high quality information and analysis provided by RISE.

Outcomes of the Capstone at Kean University

Against Ramapo's some 35 capstone projects, Kean has tackled eight topics since 2014. Both institutions involve clients for projects. At Kean, there are generally two simultaneous projects a year, each with one or two teams, although one large project accommodating multiple teams has also been carried out. For returning clients, projects have been carried out based on products from the previous year's capstone.

For example, a nonprofit called Groundwork Elizabeth has served as a Kean client for three consecutive years. Students have worked to create and manage an organic farm under their auspices located on the Kean campus. In the first year, permaculture as well as water and soil management plans were developed in addition to a crop rotation schedule, a medicinal plant garden, and educational curricula. Parts of the permaculture plan were included in a Groundwork Elizabeth proposal to the New Jersey Department of Environmental Protection, which awarded a $250,000 grant for the More P.E.A.S. (Permaculture Education in Agricultural Systems) initiative, promoting key tenets of permaculture: environmental stewardship, social and cultural revitalization, reinvestment, and recycling (Shebitz et al. 2017b). In the second and third year of having Groundwork Elizabeth as a client, the student teams implemented parts of this grant by furthering the development of the permaculture plan, creating a mobile classroom, an outdoor educational trail network, and lesson plans.

At Kean, once the final project is delivered to the client, students are asked to participate in an online survey to assess their learning outcomes and degree of satisfaction with the course structure and experience. In every assessment, the great majority of students report that the experiential learning component of the capstone is "Very effective" in teaching them how to apply environmental principles in real-life situations and that it is "Much better" than the more typical structure of the capstones that involve simply writing a major research paper and giving an individual oral presentation. The area that students tend to rank as most important is the increase in their ability to work effectively as part of a team. In addition, over 80% of students who have completed the surveys report that the practicum greatly influences their career objectives.

The students' roles in the capstone projects feature prominently on their resumes and have indeed contributed to their placement in graduate programs and careers. For example, the project manager for the farm management plan development on behalf of Groundwork Elizabeth landed a job at a local energy company as a project manager; the student who developed the water management plan received a position in water resources management.

Similar outcomes have been found at Ramapo, where close to a thousand students have participated in the capstone. It is remarkable to witness the immediate transfer of skills gained through the course into career building opportunities.

Reflection and Conclusion

NACE (2013) has claimed that successful employment upon graduation is often dependent on professional skills involving team work, problem solving, decision making, organization, and communication. In the capstone framework implemented at both Ramapo College and Kean University, students are given the opportunity to develop this essential skillset. Each semester, managing team dynamic issues represents a substantial task for both instructors and students. Yet, this challenge pays off in the evident growth of confidence and skills for the participating students, the success of the project for the clients, and ultimately the ability of students to move into positions where group and organizational skills are a significant asset.

Both Ramapo and Kean are located within the most densely populated state in the United States, where many residents have become disconnected from their environment. The evident anthropogenic impact on the land creates limitless opportunities for projects based on EIA and sustainability design and implementation. The connection between undergraduates and their ecological and neighborhood communities provides a foundation upon which long-term partnerships, skillsets, and careers are built.

While the fit of the capstone approach described in this chapter to Environmental Studies is self-evident, its employment in Ethnobiology and Human Ecology deserves some reflection. Given that it combines Biology and Anthropology around the constructs of culture and history, Ethnobiology is in a unique interdisciplinary position to bridge the sciences and humanities (Wolverton et al. 2011). More broadly, this capstone approach meets the prevailing view of Conservation practitioners that scientists must be socially involved, interdisciplinary in their approach, and able to work on local, regional, and global levels (Rozzi et al. 2006). There is also a growing recognition that Traditional Environmental Knowledge can deeply inform our scientific and cultural perspectives (Turner and Berkes 2006). This combination of western and Indigenous ecological knowledge was evident in the Ramapo College capstone collaboration with the Turtle Clan of the Ramapough-Lenape Tribe described above.

Since NEPA, there has been an increasing demand for scientists to be involved with environmental decision makers and educators (Rozzi et al. 2006). We have shown that capstone projects bridging science, policy, and the public not only positively benefit students by preparing them for the real world, but additionally benefit the greater community. This approach to the professional empowerment of upper-level undergraduates working with community partners is applicable to multiple fields and projects. While we present an overview of projects that have been conducted at our own institutions, there is no shortage of opportunities for educators from throughout the world to develop a pedagogy that relies on collaborative learning and that crosses the college-community boundary. Indeed, we believe that the lessons instilled through a capstone course will have the most meaningful and long-lasting effects through this approach.

References Cited

Appleby, K.M., E. Foster, and S. Kamusoko. 2016. Full Speed Ahead: Using a Senior Capstone Course to Facilitate Students' Professional Transition. *Journal of Physical Education, Recreation & Dance* 87(3):16–21.

Brewer, C., and D. Smith. 2011. Vision and Change in Undergraduate Biology Education: A Call To Action. Final Report of a National Conference Organized by the American Association for the Advancement of Science. Washington D.C. URL: https://live-visionandchange.pantheonsite.io/wp-content/uploads/2013/11/aaas-VISchange-web1113.pdf. Accessed on March 24, 2018.

Committee on Undergraduate Biology Education to Prepare Research Scientists for the 21st Century; Board on Life Sciences; Division on Earth and Life Studies; National Research Council. 2003. *Bio2010: Transforming Undergraduate Education for Future Research Biologists*. National Academy Press, Washington, DC.

Edelstein, M.R. 2004a. Sustaining Sustainability. In *Strategies for Sustainability: Stories from the Ivory Tower,* edited by Geoff Chase and Peggy Barlett, pp. 271–292. MIT Press, Boston.

Edelstein, M.R. 2004b. *Contaminated Communities: Coping with Residential Toxic Exposure,* 2nd edition. Westview Press/Perseus Books (now Taylor and Francis), Boulder, CO.

Edelstein, M.R. 2009. Ecological Literacy in the Undergraduate Curriculum: Ramapo College of New Jersey's Pioneer Sustainability Curricular Infusion Project. *The Place for Sustainability in Higher Education* Special Issue: *Thresholds in Education* 35(1–2):23–31.

Franchetti, M., and S.S. Ariss. 2016. The Implementation of Senior Design Capstone Projects Combining Engineering and Business Students. *Journal of STEM Education* 17(4):29–34.

Franchetti, M., M. Samir Hefzy, M. Pourazady, and C. Smallman. 2012. Framework for Implementing Engineering Senior Design Capstone Courses and Design Clinics. *Journal of STEM Education* 13(3):30–45.

Grahe, J.E., and R.C. Hauhart. 2013. Describing Typical Capstone Course Experiences from a National Random Sample. *Society for the Teaching of Psychology* 40(4):281–287.

Hauhart, R.C., and J.E. Grahe. 2012. A National Survey of Sociology and Psychology Senior Capstone Courses. *Teaching Sociology* 40:227–241. DOI:10.1177/0092055X12331715.

Kuh, G.D. 2008. *High-impact Educational Practices: What They Are, Who Has Access to Them, Why They Matter.* AAC&U, Washington, DC.

Latif, J., S.I. Sam, and G.A. Smith. 2015. Capstone Interdisciplinary Team Project: A Requirement for the MS in Sustainability Degree. *International Journal of Sustainability in Higher Education* 16:187–199.

Levi, D. 2010. *Group Dynamics in Teams.* Sage Publications, Thousand Oaks, CA.

Marriott, B.B. 1997. *Environmental Impact Assessment, a Practical Guide.* McGraw Hill, New York.

National Association of Colleges and Employers (NACE). 2013. Job Outlook: The Candidate Skills/Qualities Employers Want [web page]. URL: https://www.naceweb.org/about-us/press/2017/the-key-attributes-employers-seek-on-students-resumes/. Accessed on September 19, 2021.

Rozzi, R., F. Massardo, C. Anderson, K. Heidinger, and J.A. Silander, Jr. 2006. Ten Principles for Biocultural Conservation at the Southern Tip of the Americas: The Approach of the Omora Ethnobotanical Park. *Ecology and Society* 11(1):43. [online] URL: http://www.ecologyandsociety.org/vol11/iss1/art43/.

Shebitz. D.J., S. Capozzi, and J.P. Albaum. 2017b. Planting More than Just Veggies: Student-Centered Plans for a Sustainable Urban Farm. *Journal of Sustainability Education* 14 [online] URL: http://www.susted.com/wordpress/content/planting-more-than-just-veggies-student-created-plans-for-a-sustainable-urban-farm_2017_06/. Accessed on September 19, 2021.

Shebitz, D.J., W. Eaton, and J. Ha. 2017a. Undergraduate Learning from the Ground Up: Linking Belowground and Aboveground Diversity in Costa Rica. *CUR Quarterly* 37(3):11–17.

Turner, N.J., and F. Berkes. 2006. Coming to Understanding: Developing Conservation through Incremental Learning in the Pacific Northwest. *Human Ecology* 34:495–513. DOI 10.1007/s10745-006-9042-0.

Wolverton, S., C.R. Randklev, and A. Barker. 2011. Ethnobiology as a Bridge between Science and Ethics: An Applied Paleozoological Perspective. In *Ethnobiology,* edited by E. Anderson, D. Pearsall, E. Hunn, and N. Turner, pp. 115–132. Wiley-Blackwell, Hobeken, NJ.

APPENDIX 1.

Elk Hide Tanning Activity

Fidji Gendron[1], Vincent Ziffle[1], and Nametha Srikanthan[2]

This activity is an adaptation on traditionally tanning a hide in a laboratory setting. We used an elk hide (Figure A1.1) but other types of hide can be used. The steps are not necessarily the same in the laboratory as they would be in the open environment, utilizing the tools Mother Nature provides. A female Elder taught the First Nations traditional teachings for the hide tanning and this workshop was the first time the Elder had ever done this activity indoors.

Figure A1.1. An elk in the wilderness.

Normally, hide tanning would take place outdoors. After the hide was skinned it would be brought to a river and staked to its edge. The rushing water would rinse the fur and animals would come and eat the fat and other loose tissues. The hide would then be mounted on tree branches or over a log or tree trunk for scraping the flesh off of the hide surface.

1 Program of Indigenous Knowledge and Science, First Nations University of Canada, 1 First Nations Way, Regina, Saskatchewan, Canada, S4S 7K2

2 College of Nursing, University of Saskatchewan.

Materials

- Hide
- Tobacco offering for the Elder
- Two 76-L buckets
- Hose for filling buckets
- Plastic tarp
- Gloves
- Safety glasses
- Aprons (water proof)
- Shoe covers
- Towels
- Ash from a fire (in a 925 g coffee tin)
- Fleshing knives, scraping tools, scalpels/knives (Figure A1.2)
- Traditional tools such as elk rib bones and sharp rocks
- Wooden frame for stretching hide
- Thick rope
- Corn oil
- Paintbrush to oil the dry hide

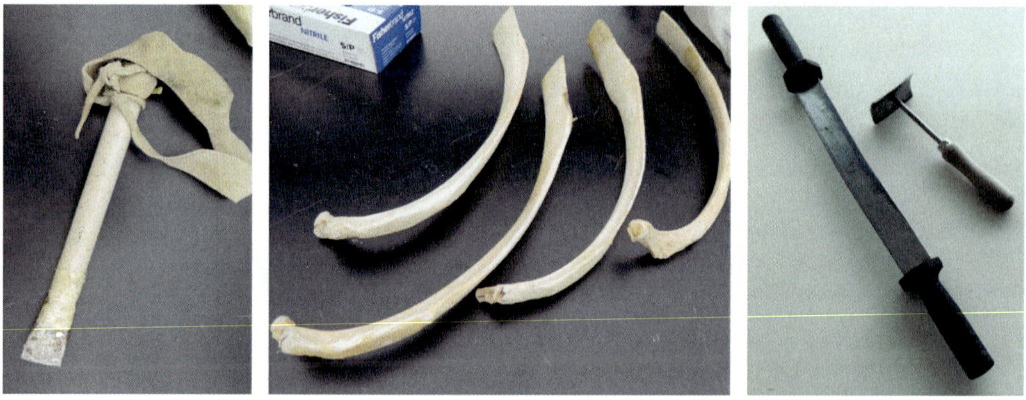

Figure A1.2. Left: Man made scraping tool made from metal, wood, and leather. Middle: Elk rib bones used as scraping tools. Right: Fleshing knife and scraping tool.

Preparing the Hide

For our workshops, the hide was stored in a deep freezer from the time it was skinned from the elk to the time of preparation. It was thawed in the refrigerator before use to minimize spoilage.

Plastic wrap or tarp was laid out in the working area. Protective gloves and safety glasses were worn. Aprons and shoe covers were used to prevent spreading of hair and other debris throughout the building space. If you think the hide you are working with is too big, cut the large hide in two, preferably before the hide has been soaked, using a sharp knife.

The hide was left to thaw in a refrigerator for two days before the activity. Once the hide was thawed, a large empty bucket was filled with clean water. We used hoses connected to the sink tap to fill up the bucket (Figure A1.3). The hide was dunked into the water repeatedly like a cloth. The hide was transferred to a second bucket filled with clean water and submerged. Dirty water from the first bucket was poured out and refilled with clean water. Transferring the hide to clean buckets of water was repeated several times until the hide was fully rinsed and the water was clear (Figure A1.4).

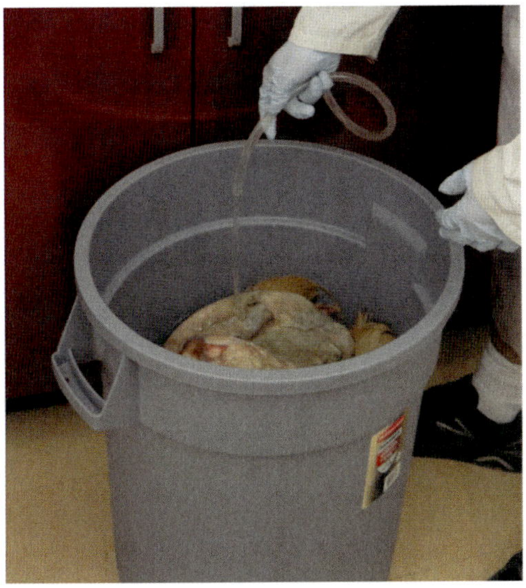

Figure A1.3. Filling the large bucket with clean water with the hide inside.

Figure A1.4. Filling the second bucket with clean water so the hide can be transferred for another rinse.

Wringing out the hide took two or more people (Figures A1.5 and A1.6). The hide can be slippery and, depending on its size, heavy. After, the hide was placed flesh-side-up on a large towel and left to dry. Salt can be used at this time to draw out water from the hide. It is recommended that the hide dries as much as possible.

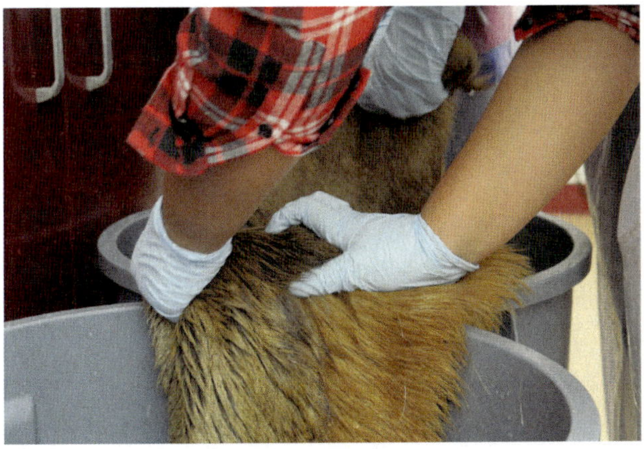

Figure A1.5. Wringing out the elk hide.

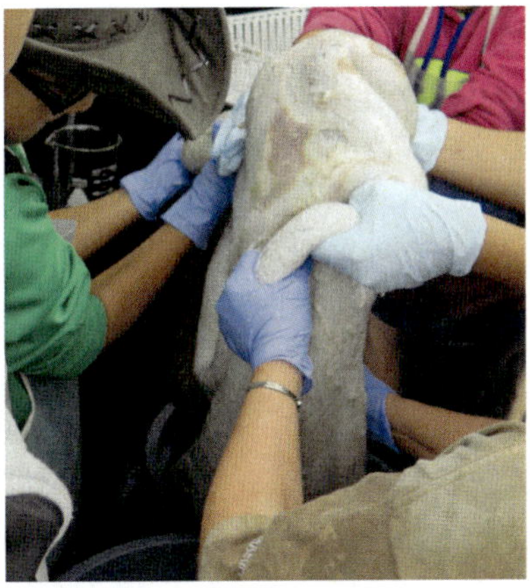

Figure A1.6. Multiple people wringing out the elk hide.

A wood frame was built during the workshop to tie up the hide for stretching. The edges of the hide were hole-punched for the rope to lace through and the hide was placed in the center of the frame and stretched out towards the edges of the frame. Cutting holes in hide led to ripping and tearing so we used a sharpened cylinder to make holes. Figures A1.7 and A1.8 show our first frame. This one was too large so we made a smaller frame after that.

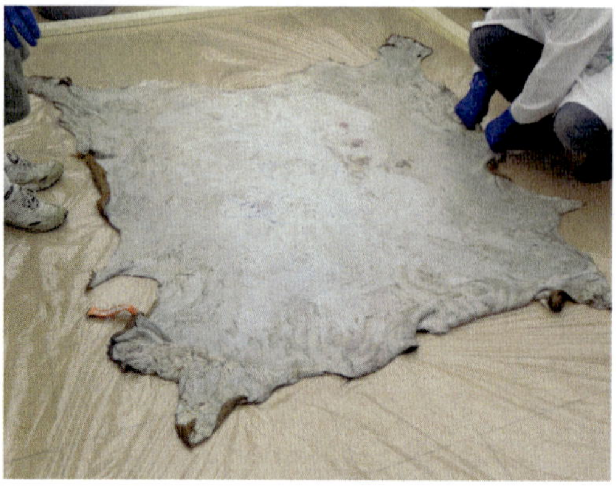

Figure A1.7. Placing the hide in the center to be stretched on the frame.

Figure A1.8. Lacing a rope through the holes in the hide to be stretched on the frame.

After the hide was stretched tightly the first scraping was done on the flesh side of the hide. Traditional tools such as sharp rocks and rib bones were used as well as other tools such as a long fleshing knife, scraping tools, and scalpels. An alternative way of scraping the hide was also done without putting the hide on the frame and it required some people stretching the hide while others were scraping—this could be done on a table or on the floor (Figures A1.9 and A1.10). To remove flesh efficiently, fast and repetitive movements are recommended. Put the sharp edge of the tool on the hide and drag it towards you with a lot of pressure and force

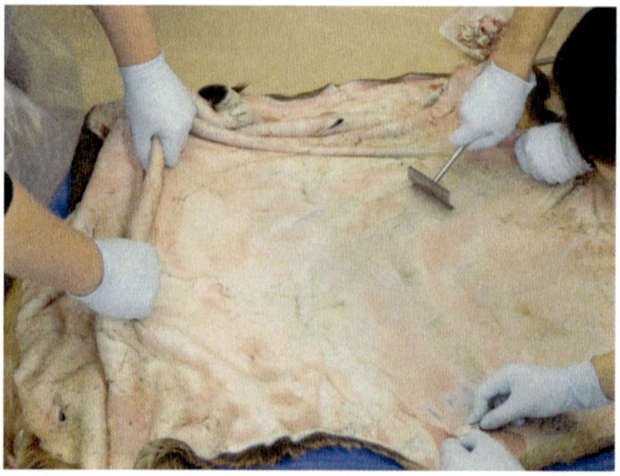

Figure A1.9. Using scraping tools to take the flesh off the elk hide.

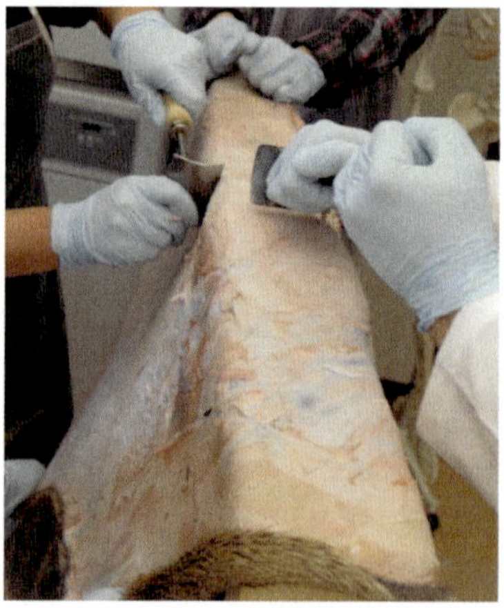

Figure A1.10. People stretching the elk hide while others scraping.

for a short distance (5 cm) and lift the tool up and place it back on the starting position and repeat the motion. Eventually the piece of flesh you are working on will begin to peel back and then it can be ripped off by hand or cut off. Care must be taken to avoid ripping of the hide.

After all flesh was scrapped off, then ash water was prepared. Wood ash in water was used to loosen the fur fibers from the hide and it helped to take the fur off easier. A 925 g coffee tin full of ash from a fire was sifted with a strainer as large pieces of charcoal could stain the hide (Figures A1.11–A1.13). The larger pieces of charcoal were crushed into finer pieces. All the ash was eventually poured into the large bucket of water. The fur was left to soak in ash water overnight. Large rocks were used to weigh down the hide so it was completely submerged in the ash water (Figure A1.14).

Figure A1.11. Filling up the bucket with water in preparation for the ash.

Figure A1.12. Sifting the coffee tin full of wood ash.

Figure A1.13. Larger pieces of ash are prevented from falling in the water.

Figure A1.14. Elk hide is held down by rocks to be completely submerged in the ash water.

After washing the hide again several times in buckets to get rid of the ash and debris, the hide was again stretched tightly in the frame and it was scraped once more, this time to remove the fur (Figures A1.15 and A1.16). Rocks and elk rib bones were used (Figures A1.17–A1.19).

Figure A1.15. Elk hide being laced on to the frame.

Figure A1.16. Elk hide almost completely laced and stretched on the frame.

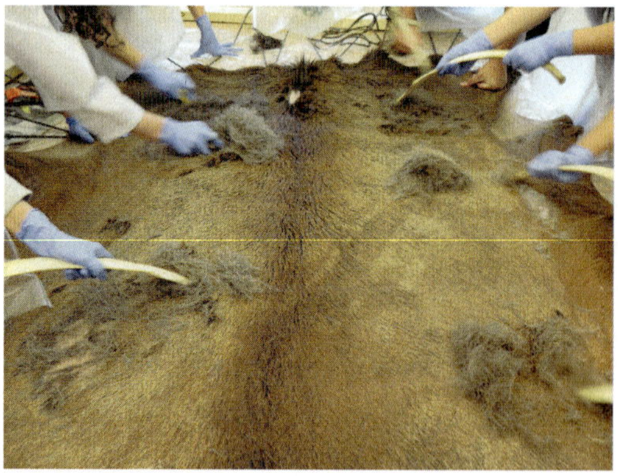

Figure A1.17. Multiple people using elk rib bones to scrape off the fur.

Figure A1.18. Close up of scraping fur off the elk hide with elk rib bone.

Figure A1.19. Close up of multiple people scraping fur off elk hide.

After the fur was completely removed from the hide it was left on the frame to dry out. The hide can be oiled with bear grease or corn oil using a paintbrush (Figures A1.20–A1.22).

Rawhide can be used to make drums or rattles. To use the hide for making drums, it must be soaked again and stretched over the frame of the drum and let to dry.

One elk hide can make up to six pairs of moccasins. To use the hide for making coats, more processing of the hide is needed. Sinew from the hump of an elk would be used to sew together hide when making a coat.

Figure A1.20. Commercial oil with paint brush and beaker container.

Figure A1.21. Oiling the elk hide with commercial oil and paint brush.

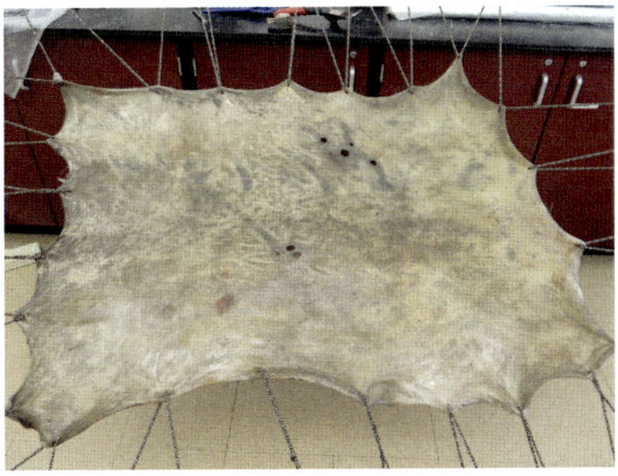
Figure A1.22. Elk hide completely stretched and freshly oiled.

Appendix 2.

Sage Hand Salve—Traditional Plant Uses Activity

Fidji Gendron[1], Vincent Ziffle[1], and Nametha Srikanthan[2]

This is an activity guided by Elder. In this activity you will learn how to prepare sage hand salve.

Materials

- 2 cups (500 ml) of oil, olive or coconut oil (Bear grease was traditionally used)
- 1/2 cup (125 ml) crushed sage (Figures A2.1 and A2.2)
- 1/4 cup = 30 g of beeswax
- Container to boil the sage (a beaker)
- Spoon to stir the oil and sage infusion
- Hot plate
- Cheesecloth
- Small containers

Figure A2.1. Prairie sage.

Figure A2.2. Pasture sage.

1 Program of Indigenous Knowledge and Science, First Nations University of Canada, 1 First Nations Way, Regina, Saskatchewan, Canada, S4S 7K2
2 College of Nursing, University of Saskatchewan.

Directions

1. To infuse the oil with the sage place in a beaker on a hot plate.
2. Heat on low heat for 2 and 1/2 hours.
3. Stir frequently.
4. When oil is green strain it with cheesecloth and let it drip then squeeze out.
5. Discard sage residue.
6. Place green oil on the hot plate to reheat on low and add the beeswax, stirring until melted.
7. Pour into jars.

Preparing the Salve

When harvesting the sage, never pull it out by the root—the root must remain in the ground for it to grow back. Cleaning the sage is done by taking out all the grasses and dirt from the bundles and rinsing with water. The sage is then left to dry for one week.

Prairie sage (*Artemisia ludoviciana*) is typically used for salves but pasture sage (*Artemisia frigida*) can be used as well (Figures A2.1 and A2.2). Prairie sage is native to Saskatchewan and it is found in semi-arid climates such as prairies and it is usually harvested from July to September. The Elder we worked with referred to prairie sage as female sage and pasture sage as male sage and said they should not be mixed or ground together when making the salve. Prairie sage crushed well when it is dried and pasture sage gets fluffy. It is best to use a coffee grinder for pasture sage. Dried sage is finely ground with a rock, a pestle and mortar, or a coffee grinder (Figure A2.3). All the parts of the plant can be ground and used.

Figure A2.3. Grinding the sage in coffee grinders.

Oil was poured into a container and heated on a hot plate—in the laboratory a beaker was used for the container. Bear grease is traditionally used as the oil—both bear grease and sage are medicines, but olive oil or coconut oil can also be used. If coconut oil is used it may overpower the scent of sage. Once the oil was hot, the sage was added to the oil and boiled for two hours. The oil can be infused with other aromatic flowers as well.

The olive oil started out bright yellow but when it is infused with sage it is green. The sage and oil infusion were strained with cheesecloth over the beaker (Figure A2.4). The oil and sage extract ended up in the beaker and the sage residue was filtered out (Figure A2.5). The residue of sage can be used as an exfoliator for skin and face—it is good for treating acne.

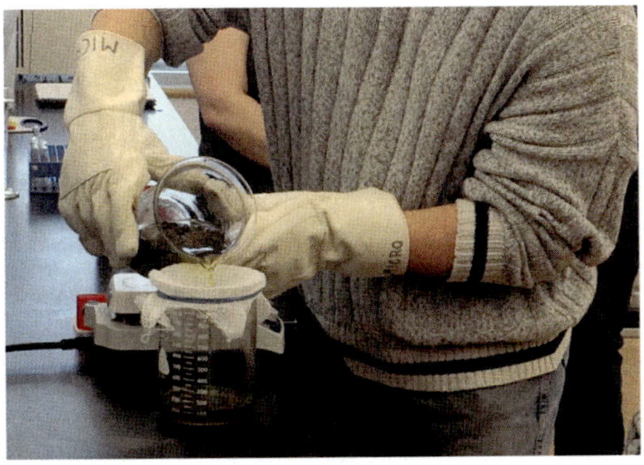

Figure A2.4. Straining sage and oil infusion over cheesecloth and beaker.

Figure A2.5. Squeezing the infusion out of the cheesecloth and into the beaker.

The oil was strained quickly to keep its hot temperature and placed on the hot plate again. Pure beeswax was added to the hot oil and with gentle stirring it melted. The salve will turn rancid if there is any honey left on the beeswax, so it was cleaned out very well.

After all the beeswax was melted in the oil, a scoop was used when the hot oil was poured into several small closeable containers (Figures A2.6 and A2.7). These are the final containers for the salve. When the oil cooled down it solidified because beeswax melted in the oil is solid at room temperature.

During the activity, the Elder shared stories about sage and each student had its own hand salve sample at the end (Figure A2.8).

Figure A2.6. Scoping the hot sage oil into a small closable container.

Figure A2.7. Three small closeable containers.

Figure A2.8. The elder with completed sage hand salves.

Appendix 3.

Porcupine Quill Bookmark Activity

Fidji Gendron[1], Vincent Ziffle[1], and Nametha Srikanthan[2]

Source: http://maxpixel.freegreatpicture.com/Porcupine-Canada-Animal-Prickly-Canadian-Porcupine-381299

This is an activity guided by Elder. In this activity you will learn how to make a bookmark featuring a design made with dyed porcupine quills. The porcupine quills must be dyed prior to the activity. The porcupine used in this activity booklet was a roadkill that was donated to the University.

Traditional Knowledge Shared by Elder

Traditionally, quills were obtained from a live porcupine. A salt lick was often used to lure the porcupine and a gunny sack (essentially a sack made of burlap material) was thrown onto the porcupine. Before the porcupine ran away, the gunny sack was pulled off the porcupine and the quills would remain stuck to the gunny sack. Elder said, "There is no reason to kill the

1 Program of Indigenous Knowledge and Science, First Nations University of Canada, 1 First Nations Way, Regina, Saskatchewan, Canada, S4S 7K2

2 College of Nursing, University of Saskatchewan.

porcupine for its quills because it is not good to kill your store." Since the porcupine can grow back its quills again and again, if the porcupine was dead, no more quills could be obtained from it. Also, if quills are obtained from a dead porcupine they may get damaged from tools such as pliers that are used to pull them out.

The quills of the porcupine are like hair, and like hair they can be dyed. When using natural dyes, i.e., dye from berries, boiled black spruce was traditionally added. The colour can be picked up by the quill better with the black spruce. Artificial dyes can also be used.

The older the porcupine, the longer the quills are, and they become oily. According to Elder this oil is dark and rancid. Traditional quillers—those who work with the quills—used to flatten the quills out with their teeth and you can tell someone was a quiller when they had stained lips from the quill oil and from the dyes.

Before beads were used for making intricate designs, quills were used for this purpose. The quills were traditionally stitched on to cloth, leather, or birch bark. The birch bark box in the photo below was made by Elder (Figure A3.1).

Figure A3.1. A birch box decorated with porcupine quills, with some of them dyed pink and blue.

Bookmark Making Procedure

Materials

- Porcupine quills
- Rit liquid dye (from craft store)
- Rubber gloves
- Beakers
- Hot plate
- Thermometer
- Glass stirrer
- Strainer
- Scissors
- Fabric (canvas)
- Ruler
- Pencil
- Thread
- Embroidery needles (from craft store)

Dying the Quills: Time ~1 hour

Wear rubber gloves. Line up quills to ensure you are using quills that are similar in width and length (Figure A3.2). Heat 500 ml of water in a 1 L beaker over a hot plate set to 300°C. As per the directions on Rit liquid dye bottle (Figure A3.3), one bottle (236 ml) should be used for 1 kg of material to be dyed. Weigh the selected quills and put in formula:

236 ml/kg x weight (kg) = amount of dye (ml)

Figure A3.2. Lining up the quills.

Figure A3.3. Rit liquid dye.

For example, 7 g of quills: 236 ml/kg x 0.007 kg = ~1.7 ml of dye added to 500 ml of water. Check the temperature of the water with the thermometer, when it is ~80°C, add the quills to the water and dye solution, and start the timer for 30 minutes (according to instructions on Rit liquid dye bottle). Stir the quills frequently using a glass stirrer to ensure all quills are evenly coated in the dye. After 30 minutes, use a metal strainer to strain the quills and rinse under running water (Figure A3.4). If the holes of the strainer are big, place a paper towel in first before straining the quills so they do not slip through holes. When the water runs clear, leave the quills to dry on a paper towel (Figure A3.5).

Figure A3.4. Straining the quills.

Figure A3.5. Leaving quills to dry.

Porcupine Quill Preparation: Time ~15 minutes

Pick out five to ten matching quills in length and width. The small pointy white end of the quill should be snipped off completely with scissors (Figure A3.6). When snipping off pointy ends, do not snip off the entire black part because you will lose the signature look of the quill.

Using a pencil and ruler, draw the design you will use for the bookmark on the strip of canvas (Figure A3.7). Thread the embroidery needle with white thread or coloured embroidery thread (Figure A3.8) but do not knot the end—this is for aesthetic purposes—knots make it obvious that it is the beginning or end of a design. The amount of thread needed varies depending on the type of design but this simple design only requires approximately an arm's length (Figure A3.9).

Figure A3.6. The entire white tip shown in this figure will be snipped.

Figure A3.7. A simple design for the canvas bookmark.

Figure A3.8. Multiple colors of embroidery thread.

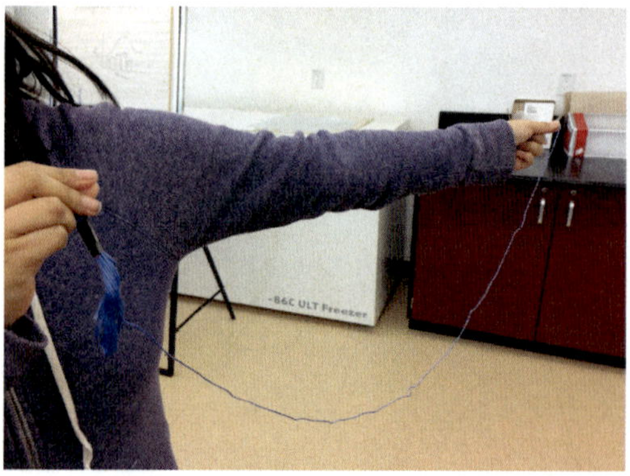

Figure A3.9. An arm's length of thread can be used.

Straight-line Puff Pattern: Time ~30–45 minutes

Line up the first quill along the straight line on the canvas. Push the needle and thread up from under the canvas slightly to the right of your quill then wrap the thread around the quill a couple times (pulling tight) then push the needle back down into the same spot from which it came up. Do the first stitch twice because it anchors the quill on to the canvas (Figure A3.10). The next stitch will be spaced out from the first and the length in spacing depends on the look you want to achieve but they should be even throughout (Figure A3.11). For each stitch, you will come up from under the canvas (always on the same side of the quill), wrap around the quill, come down in same entry spot, and pull the thread tightly. The stitches will flatten the quills, creating the "puff" effect when the stitch is tightened (Figure A3.12). When adding the next quill, place it slightly underneath the previous one and make sure they are going in the same direction—have the black tip overlapping the white tip of the next quill. Use a stitch to wrap them together tightly, while making sure the stitch matches all previous stitches. Continue this procedure until your bookmark design is complete (Figure A3.13). Do the last stitch twice (similar to the first stitch) so that it is anchored to the canvas.

Figure A3.10. The first stitch anchors the quill.

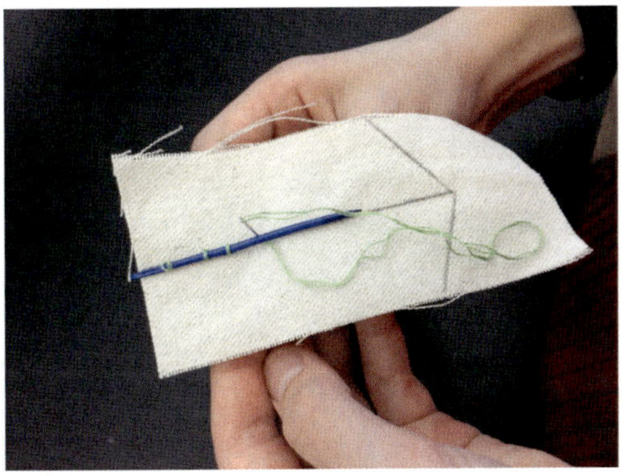

Figure A3.11. Each stitch is spaced out as evenly as possible.

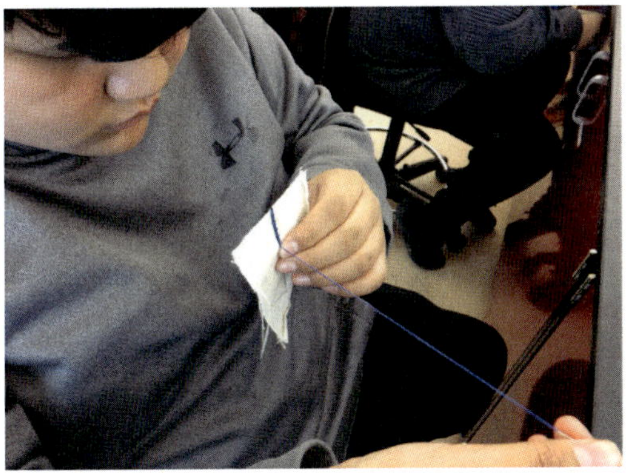

Figure A3.12. The stitch is tightened.

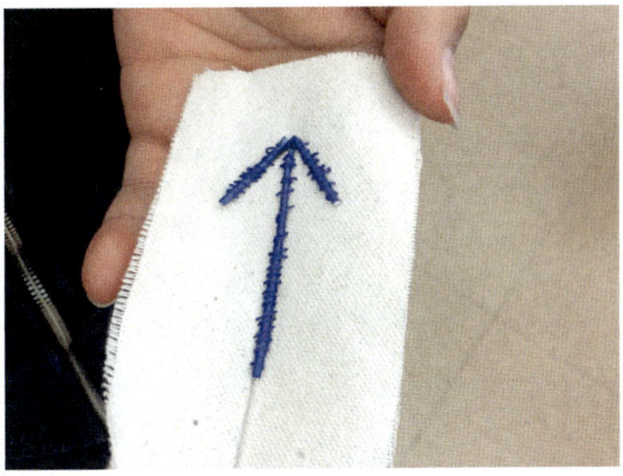

Figure A3.13. A partially finished arrow design.

Appendix 4.

Lesson 1: Culture and Ecology in Mesoamerica

Amanda M. Thiel[1] and Marsha B. Quinlan[1]

While definitions vary, for this lesson we will consider Mesoamerica to be a combination of Mexico and all of the countries of Central America (Belize, Guatemala, El Salvador, Nicaragua, Honduras, Costa Rica and Panama; Figure A4.1). Mesoamerica means "middle America" and is thus aptly named for its location between the US in North America and Colombia in South America.

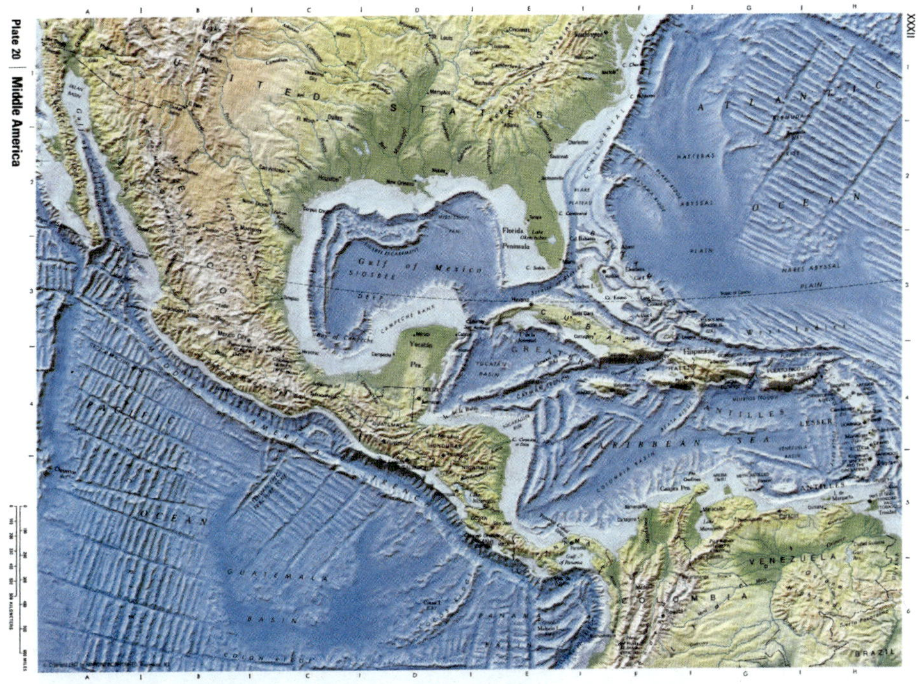

Figure A4.1. Map of Mesoamerica (Hammond 1987).

1 Washington State University

Ecology and Culture

Mesoamerica is a very diverse area, both ecologically and culturally. Ecologically speaking, it is a merging of two major continents (North and South America), thus giving it a lot of different ecoregions, or ecological zones and mini-climates. The entire area contains desert, mountains, valleys, beaches, tropical wet forest, and tropical dry forest. The elevation reaches from sea level to volcanoes of 3,048 m (Skidmore et al. 2014).

Levels of development and infrastructure are quite varied throughout the area, from virtually none, to major urban centers like Mexico City and Panama City (e.g., Figures A4.2 and A4.3). Urban centers are home to the majority of the population in each country. For example, one in five Costa Ricans live in the capital, San Jose, and one in four Mexicans (21 million people) live in Mexico City (CIA 2019). People living rurally migrate to cities in search of jobs.

Figure A4.2. Photo of rural Guatemalan thatched-roof home (Amanda M. Thiel).

Figure A4.3. Photo of Panama City (Mariordo 2013).

The diversity of environments allows for many different types of land use and modes of subsistence, including hunting-and-gathering, pastoralism, horticulture, and industrial agriculture. Slash and burn horticulture is common and has been practiced for millennia (Ford and Nigh 2016; McKillop 2004). It centers around the triad of crops known as the *milpa*—corn, beans, and squash. Corn in local cuisine is often prepared into tortillas, empanadas, arepas, or tamales. Most large-scale commercial farms are huge tracts of land owned by one person, or by a company. The most common crops are bananas, coffee, sugar cane, African palm (for oil), citrus, and cotton. Plantations have a reputation for deplorable social and environmental impacts (Prieto-Carron 2006; Skidmore et al. 2014).

Watch this short video about the process of food production from farm (*milpa*) to table (*mesa*): https://vimeo.com/205895612.

Culturally, Mesoamerica is home to many different ethnic groups. The largest ethnicity is mestizo (mixed European and indigenous ancestry). Other groups include indigenous, mulato (mestizo and black/African descent), white, and Garifuna (descendants of African, European, Island Carib and Arawakan people). Mexico and Guatemala have the highest percentages of indigenous people (28% and 39% respectively; CIA 2019). Almost all Mesoamerican countries share a Spanish colonial heritage, including mandatory adoption of the Spanish language, as well as architectural elements and cultural practices. It is, in great part, this shared linguistic and colonial patrimony that unites an otherwise culturally and ecologically diverse region. The official language in Mesoamerican countries is Spanish, with the exception of Belize, whose primary and official language is English. For indigenous Mesoamericans, however, their mother tongue is usually an indigenous language (e.g., Nahuatl, Zapotec, Mixtec, Miskito, Kuna, or one of several Maya languages) and bilingualism with Spanish is common (CIA 2019; Skidmore et al. 2014).

Catholicism has long been the most common major religion throughout Mesoamerica (and Latin America generally), although this is often syncretized, or blended, with indigenous religious beliefs and practices among indigenous Mesoamericans. In the last couple of decades there has been a growing trend toward Evangelism, especially in Guatemala, El Salvador, and Nicaragua (CIA 2019; Wilson 1995).

A Brief History of the Area

Mesoamerica has a rich cultural prehistory (known as the pre-Columbian era [link to https://www.smith.edu/vistas/vistas_web/units/surv_precol.htm]) with numerous indigenous groups forming advanced civilizations including architectural, astronomic, artistic and other cultural achievements (McKillop 2004; Figure A4.4).

The Spanish colonial era began in Mexico in the early 1500s and eventually spread throughout Mesoamerica and neighboring regions. In the early 1800s, independence movements began, over time giving way to various permutations of nationhood, at one point including the Federal Republic of Central America, and eventually, independent nation states as we know them today (Skidmore et al 2014).

Figure A4.4. Chichen Itza pyramid, in Yucatan, Mexico. Perhaps the most famous Mayan city and structure dating to 600 AD (Amanda M. Thiel).

A short animated history of the area: https://www.youtube.com/watch?v=bUoypQ0ykwI.

Much of the history of Mesoamerica has been characterized by unequal land distribution. This began, of course, with the usurpation of indigenous Mesoamerican's land, lives and life-ways by the Spanish upon conquest, and the pattern of inequality continues to this day. From more-or-less the late 1800s to the 1930s, political powers transferred land ownership from communal systems to private ones, and opened up land to foreign investment in an effort to improve commerce. The privatization of land in the 1940s and 50s by foreign agricultural investors, such as the US-based United Fruit Company and German coffee and sugar plantations, meant the exploitation of local people throughout the entire region (Skidmore et al. 2014).

The US had major interest in Central America, and "through trade, investment, invasion, and diplomacy... exerted extraordinary influence over trends and events in this area for more than 100 years" (Skidmore et al. 2014:79). The US controlled the Panama Canal—a strategic political and geographic asset—from the early 1900s until 1977. In the 1950s, the US felt increasingly threatened by Central American and Caribbean leaders' ties with communist Europe. As the Cold War broke out, the US militarized forces throughout Latin America, hoping to avoid the spread of communism. Latin America's response was to develop its own various guerrilla insurgencies. Horrendous violence ensued, using such methods as abductions, bombings, assassinations, and torture. Over 90% of the violence between the 1960s and 1990s was perpetrated by US-backed Central American governments, and the remaining was by guerrilla forces mostly made up of peasants, indigenous, and rural poor (e.g., the Sandinistas in Nicaragua, and the Farabundo Martí National Liberation Front in El Salvador). Civil wars broke out across Central America as political tensions waxed: El Salvador suffered a 10-year civil war, and Guatemala, a 36-year civil war. Hundreds of thousands of people died or "disappeared" in these conflicts (Skidmore et al. 2014). These events have left Central America with a legacy of land-use tensions, inequality, and violence.

Current Issues

Immigration

Illegal/undocumented immigration from Mexico and Central American countries to the US is a hot political topic that has persisted, and perhaps heightened, over recent decades. There has been a sharp rise in illegal immigration rates from Mexico and Central American countries since the 1980s, despite increasingly aggressive efforts to curb the flow of immigrants (MPI 2018; Skidmore et al. 2014).

Why do Mesoamericans immigrate to the US illegally? Originally, the reasons were triple fold: violence from civil wars, political instability, and poverty. Over time, family reunification and natural disasters became prominent reasons, as well. Gang violence, high homicide rates, and poverty continue to be some of the main factors today, especially from the countries of the "Northern Triangle," formed by Guatemala, El Salvador, and Honduras. In 2015, 85% of Central American immigrants were from those three countries (Lesser and Batalova 2017). As the immediate neighbor to the US, Mexico maintains the highest number of migrants, with one out of every ten Mexican citizens now residing in the US (Alba 2010). Illegal immigrants have come to occupy important roles in the US economy, providing experienced labor at low costs, in construction, agriculture, and various services. The money they make in the US is often sent home to their families in their country of origin which helps boost the home economy (Skidmore et al. 2014).

Check out this interactive graph showing the immigrant population in the US over time based on country of origin. Select Mexico, Belize, Guatemala, Honduras, Nicaragua, Costa Rica and Panama and notice the trends. From which Mesoamerican country do the majority of immigrants originate? You may also select other countries to make comparisons with immigration from other parts of the world.

https://www.migrationpolicy.org/programs/data-hub/charts/immigrants-countries-birth-over-time?width=1000&height=850&iframe=true

Drug Trafficking

Another major issue in inter-American politics is the flow of narcotics. Illegal drugs are possibly the "most profitable source of economic activity for organized crime in the United States" (Skidmore et al. 2014:478). Most of these drugs are now produced in Latin America (cocaine from Colombia/Peru/Bolivia, marijuana from Central America and Mexico, heroine from various parts of Latin America, as well as Asia). The principal consumer is the US (although Europe and Asia have high rates, too), and it is the strength of consumer demand that ultimately causes drug trafficking. Concerns over drug-related violence has sparked numerous debates about the best way to change policy regarding drug trafficking (Skidmore et al. 2014).

To see the flow of drugs, cartels, and violence in Mexico see *The Economist* (2019): https://www.economist.com/node/21555750/mobile-frameless.

Questions for Reflection

1. What are some of the cultural features (language, subsistence, etc.) that constitute daily life in Mesoamerica?
2. How has the role of outsiders (foreign governments) historically affected and been received by Mesoamericans? Why?
3. What are some of the assets and challenges of Mesoamerican people, historically and currently?

References Cited

Alba, F. 2010. Mexico: A Crucial Crossroads [web page]. Migration Policy Institute, Washington, D.C. URL: https://www.migrationpolicy.org/article/mexico-crucial-crossroads. Accessed on July 8, 2019.

Central Intelligence Agency (CIA). 2019. The World Factbook [web page]. U.S. Central Intelligence Agency, Washington, DC. URL: https://www.cia.gov/library/publications/the-world-factbook/index.html. Accessed on July 8, 2019.

Ford, A., and R. Nigh. 2016. *The Maya Forest Garden: Eight Millennia of Sustainable Cultivation of the Tropical Woodlands*. Routledge, Philadelphia.

Hammond. 1987. Middle America [map.]. Hammond Maps, Maplewood, NJ.

Lesser, G., and J. Batalova. 2017. Central American Immigrants in the United States. *Migration Information Source* 4/5/2017 [online]. URL: https://www.migrationpolicy.org/article/central-american-immigrants-united-states-2015. Accessed on July 8, 2019.

McKillop, H.I. 2004. *The Ancient Maya: New Perspectives*. ABC-CLIO, Santa Barbara, CA.

Mariordo (Mario Roberto Durán Ortiz). 2013. Panama 08 2013 Ave Balboa 7044 - CC BY-SA 3.0 [web page]. URL https://commons.wikimedia.org/wiki/File:Panama_08_2013_Ave_Balboa_7044.JPG. Accessed on July 7, 2019.

Migration Policy Institute (MPI). 2018. Countries of Birth for U.S. Immigrants, 1960-Present. Migration Policy Institute [web page]. URL: https://www.migrationpolicy.org/programs/data-hub/charts/immigrants-countries-birth-over-time?width=1000&height=850&iframe=true. Accessed on July 5, 2019.

Prieto-Carron, M. 2006. Corporate Social Responsibility in Latin America: Chiquita, Women Banana Workers and Structural Inequalities. *The Journal of Corporate Citizenship* (21):85.

Skidmore, T.E., P.H. Smith, and J.N. Green. 2014. *Modern Latin America*, 8th edition. Oxford University Press, New York.

Appendix 5.

Lesson 2: Health and Plants among the Maya of Guatemala

Amanda M. Thiel[1] and Marsha B. Quinlan[1]

Overview

Guatemala is a Spanish-speaking country bordering Mexico to the north, Belize to the east, Honduras and El Salvador to the south, and the Pacific Ocean to the west (Figure A5.1). It is highly culturally and ecologically diverse. It is considered to have a tropical climate, but there are hot and humid lowlands as well as cooler highlands, including active volcanoes. The country is approximately the size of Pennsylvania, with close to 109,000 km^2, and a population of 17 million people (CIA 2019).

Guatemala has a high percentage of indigenous people (approximately 40% of the total population) and the highest population growth rate in Latin America. The majority belong to one of over 20 different indigenous Maya ethnic and linguistic sub-groups. Poverty, malnutrition, landlessness, unemployment, illiteracy and infectious disease risks are high and "the country's large indigenous population is disproportionately affected"

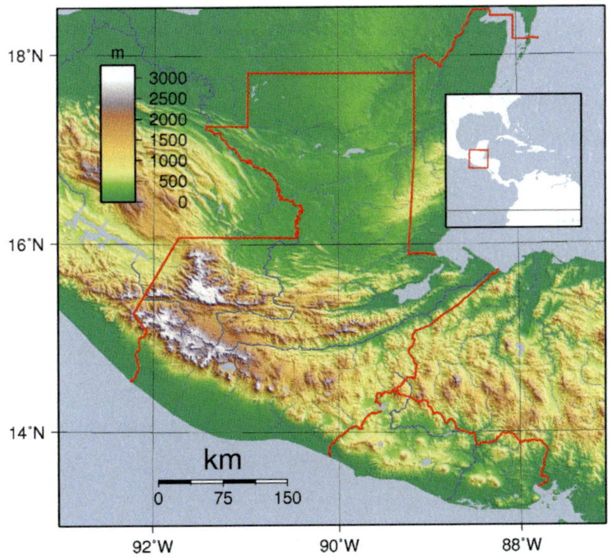

Figure A5.1. Guatemala topographic map (Sadalmelik 2007).

1 Washington State University

(CIA 2019). With globalization, many Maya people are moving away from traditional practices (of subsistence, healthcare, lifestyle, etc.) as they acculturate towards Western ideologies of development (Cosminsky 2016; Wilson 1995; Zarger 2002).

Ethnomedicine and Ethnopharmacology in Guatemala

Cross-culturally, popular sector health care, including home (and herbal) remedies, is the first step in health care (Kleinman 1980), which is especially necessary in rural, isolated and impoverished areas (Quinlan 2004), like rural Guatemala. Weller et al. (1997) find that self-treatment is the most common initial treatment-seeking behavior in Guatemala, whether via home remedies (in 56–60% of cases) or medicines from an unofficial pharmacy (30–33% of cases). Indigenous and non-indigenous people throughout Guatemala employ many medicinal plants, as they are readily available, have longstanding traditions of use, and are virtually free (Adams and Hawkins 2007; Cosminsky 2016). Much of their use, as well as other therapeutic activities, follow prescriptions of the humoral system found across Guatemala and Latin America. The humoral system requires the balancing of hot and cold qualities of food, medicine, and activities (Logan 1973; Wilson 1995), and is likely an import from Spanish colonial times and Greco-Persian-Arab medicine (Hernández Sáenz and Foster 2001).

How rural, indigenous Guatemalans care for their health with locally available botanical resources and the potential of these resources as pharmacologically active agents is important for public health in Guatemala and abroad (Michel et al. 2007). Understanding the context in which the Maya cultivate and use medicinal plants is necessary for cultural revitalization and successful integration of diverse health care models, as Western biomedicine continues to expand across rural Latin America and the world (Adams and Hawkins 2007; Caceres 1996; Cosminsky 2016).

Traditional medicine in Guatemala is threatened most by acculturation and biomedicine, and to some extent, loss of habitat for botanical species. The general trend toward Ladino[2]-ization, such as the loss of Maya languages and lifeways, coupled with the violent political and military regime of the 1980s, devalue indigenous cultural identity and practices (Comerford 1996). The growing, dominant role of biomedicine further diminishes the role of traditional health care (Cosminsky 2016). Financial resources, accessibility, and dissonance in cultural acceptability limit professional health care choices in rural Guatemala. In many cases, biomedicine is unavailable or not ideal because of the above constraints, and so services from *curanderos* (healers), *comadronas* (midwives) and other folk healers are popular choices (Adams and Hawkins 2007; Cosminsky 2016; Weller et al. 1997). Home remedies are still the first choice in health care practice.

2 In Guatemala, *Ladino* is an ethnic identity that refers to Westernized indigenous, mestizo or non-indigenous Guatemalans who speak primarily Spanish and wear Western clothes (Wilkinson 2004; Wilson 1995).

Many of the home remedies in use in Guatemalan Maya communities come from medicinal plants grown in people's own gardens. The provisioning of medicinal plants in home gardens is important because in many cases they are the first line of defense in treating common illnesses in an area that is largely lacking in access and resources to other methods of health care (Weller et al. 1997). Many of the plants grown in home gardens or sourced locally and used medicinally are beginning to be documented and tested for their medicinal actions, frequently being validated by Western scientific methods of analysis to support the plants' traditional uses (Caceres 1996; Michel et al. 2007).

Watch this short video on the cultivation and use of medicinal plants in a Maya community in Guatemala: https://www.youtube.com/watch?v=lvtGE7MwIpU (Rehder 2009). Given the threats to the cultivation, knowledge, and use of medicinal plants posed by acculturation, biomedicine, and habitat loss, it is important to find ways to maintain medicinal plant knowledge and practice in Guatemala (and elsewhere). After all, this knowledge and practice is largely what keeps the marginalized, rural and indigenous population well. Local botanical medical knowledge may have implications for global health care and response to disease, as well.

What should be done, and by whom, to maintain medicinal plant knowledge and practice for the benefit of local people and global populations?

Using the information presented within the lessons and articles, discuss what you think should be done, and by whom, to maintain medicinal plant knowledge and practice in Maya areas. Why is this important? Who benefits? What are some of the challenges and issues surrounding this effort? Frame your discussion by proposing a small development project that aims to preserve medicinal plant knowledge among Maya communities. Can this be achieved in a culturally appropriate way? If so, how? If not, why not? Use examples and conclusions drawn from the lesson materials whenever possible. For more information, see the assignment instructions.

References Cited

Adams, W., and J.P. Hawkins. 2007. *Health Care in Maya Guatemala: Confronting Medical Pluralism in a Developing Country*. University of Oklahoma Press, Norman, OK.

Caceres, A. 1996. *Plantas de Uso Medicinal en Guatemala*. Editorial Universitaria, Guatemala.

Central Intelligence Agency (CIA). 2019. *The World Factbook*. U.S. Central Intelligence Agency, Washington, DC. Available at https://www.cia.gov/library/publications/the-world-factbook/index.html. Accessed on July 8, 2019.

Comerford, S.C. 1996. Medicinal Plants of Two Mayan Healers from San Andrés, Petén, Guatemala. *Economic Botany* 50(3):327–336.

Cosminsky, S. 2016. *Midwives and Mothers: The Medicalization of Childbirth on a Guatemalan Plantation*. University of Texas Press, Austin, TX.

Hernández Sáenz, L.M., and G.M. Foster. 2001. Curers and Their Cures in Colonial New Spain and Guatemala: The Spanish Component. In *Mesoamerican Healers*, edited by B.R. Huber and A. Sandstrom, pp. 19–46. University of Texas Press, Austin, TX.

Kleinman, A. 1980. *Patients and Healers in the Context of Culture: An Exploration of the Borderland between Anthropology, Medicine, and Psychiatry.* University of California Press, Berkeley.

Logan, M. 1973. Humoral Medicine in Guatemala and Peasant Acceptance of Modern Medicine. *Human Organization* 32(4):385.

Michel, J., R.E. Duarte, J.L. Bolton, Y. Huang, A. Caceres, M. Veliz, and G.B. Mahady. 2007. Medical Potential of Plants Used by the Q'eqchi Maya of Livingston, Guatemala for the Treatment of Women's Health Complaints. *Journal of Ethnopharmacology* 114(1):92–101.

Quinlan, M.B. 2004. *From the Bush: The Frontline of Health Care in a Caribbean Village.* Wadsworth (Cengage) Publishing, Belmont, CA.

Rehder, S. 2009. Medicinal Plant Tour in San Juan La Laguna [Film]. Fundacion Salar, Guatemala. Available at https://www.youtube.com/watch?v=lvtGE7MwIpU. Accessed July 2019.

Sadalmelik. 2007. Topographic Map of Guatemala. Wikimedia Commons [web page]. URL: https://commons.wikimedia.org/wiki/File:Guatemala_Topography.png. Accessed on July 6, 2019.

Weller, S.C., T.R. Ruebush, and R.E. Klein. 1997. Predicting Treatment-Seeking Behavior in Guatemala: A Comparison of the Health Services Research and Decision–Theoretic Approaches. *Medical Anthropology Quarterly* 11(2):224–245.

Wilkinson, D. 2004. *Silence on the Mountain: Stories of Terror, Betrayal, and Forgetting in Guatemala.* Duke University Press, Durham, NC.

Wilson, R. 1995. *Maya Resurgence in Guatemala: Q'eqchi' Experiences.* University of Oklahoma Press, Norman, OK.

Zarger, R.K. 2002. Acquisition and Transmission of Subsistence Knowledge by Q'eqchi'Maya in Belize. In *Ethnobiology and Biocultural Diversity*, edited by J.R. Stepp, F.S. Wyndham, and R.K. Zarger, pp. 592–603. Univeristy of Georgia Press, Athens, GA.

Appendix 6.

Assignment: Development of Traditional Medicine and Use of Medicinal Plants in Maya Guatemala

Amanda M. Thiel[1] and Marsha B. Quinlan[1]

Overview

This assignment encourages you to articulate how and why the study of local culture and ecology helps inform appropriate development. You will write a short essay that addresses the following questions. You will then share it in the discussion board for peer feedback and discussion.

To Do

Essay:

Using the information presented within the two sets of lesson notes and the article by Michel et al. 2007, write a short essay that discusses what you think should be done, and by whom, to maintain medicinal plant knowledge and practice in Maya areas. Frame your discussion by proposing a small development project that aims to preserve medicinal plant knowledge among Maya communities.

1. Why is this important?
2. Who does it benefit?
3. What are some of the challenges and issues surrounding this effort?
4. Can this be achieved in a culturally appropriate way?
5. If so, how? If not, why not?

1 Washington State University

Use examples and conclusions drawn from the lesson materials whenever possible. In addition to the above criteria, you may wish to include:

- Cultural background
  - Describe the main features of the culture of Mesoamerica

- Identification of development needs
  - Identify what the needs are for health care in Guatemala, including cultural considerations that need to be taken into account

- Suggestion for a simple development plan
  - Based on the needs you identified, create a simple plan for development that addresses at least 1 of your identified needs and respects local culture
  - Provide a detailed rationale for how/why this development plan is appropriate and respects local culture
  - Keep your idea for development small and simple. Think realistically about one simple thing that could be implemented with a high degree of feasibility.

- Tips on structure of your writing:
  - Organize your essay into three or more paragraphs
  - Use correct grammar and spelling
  - Use a basic font, 12 point size, and double space
  - Write ~1–2 pages (double-spaced)
  - Write your essay in a word document, then copy it to the online discussion board

Peer Review:
Once you write the essay in a word document, copy and paste it into the online discussion board. Once peers' work is available, read and provide feedback on two of your classmates' entries. Provide constructive criticism on their development plan and how well they addressed each of the (numbered) points listed above. After your entry receives feedback, respond fully to the feedback that students gave of your work. Adjust your plan as needed and provide sound justification for any changes.

Reference Cited

Michel, J., R.E. Duarte, J.L. Bolton, Y. Huang, A. Caceres, M. Veliz, and G.B. Mahady. 2007. Medical Potential of Plants Used by the Q'eqchi Maya of Livingston, Guatemala for the Treatment of Women's Health Complaints. *Journal of Ethnopharmacology* 114(1):92–101.

Author Biographies

Dr. Michael R. Edelstein is the Convener of the Environmental Studies major at Ramapo College of New Jersey, where he teaches the courses Environmental Assessment, Field Project in Environmental Assessment, Environmental Psychology, World Sustainability and Sustainable Communities. An Environmental Psychologist, Edelstein has spent his career studying the psycho-social impacts of what he calls "Environmental Turbulence," which includes environmental contamination, destruction and undesired and hazardous change as well as advocating for a shift toward a sustainable society. He has a particular interest in how indigenous communities are impacted by environmental change and is currently working on a project in Hawaii on the impacts of geothermal energy on Native Hawaiians. He has written widely about his work, including the book Contaminated Communities for which he is currently preparing a third edition.

Dr. Fidji Gendron is interested in traditional uses of native plants. She works in partnership with Elders and has developed booklets on plant medicines. These teaching tools help promote native plants recognition and highlight their importance in the Indigenous culture. She also examines the bioactive chemical components in native plants. Another area of interest is how to braid Indigenous and Western Science in her Biology courses. She organizes hands-on activities in her labs under the guidance of an Elder who shares protocol and stories with students.

Dr. Denise M. Glover is a cultural and environmental anthropologist, and an ethnobiologist. Her research is centered in Southwest China with doctors and knowledge holders of traditional Tibetan medicine, with a focus on the linkages between place, identity, and medicine production. She was professor of anthropology and Asian Studies at the University of Puget Sound from 2008–2020 (and currently holds affiliate status there). She is an Associate Editor for the *Journal of Ethnobiology* (she was Society of Ethnobiology Treasurer for six years and Conference Coordinator for SoE for six years before that). When not teaching, researching, or writing/recording music, she often hangs out with horses (here pictured with a pony in Tibetan town of Rgyalthang, PRC).

Dr. William H. Heyborne received a BS degree in zoology from Southern Utah University, earned an MS in entomology from Oregon State University, and completed a PhD in biology (studying the chemical ecology of venomous snakes) from the University of Northern Colorado. Following completion of his doctoral degree, Dr. Heyborne began work at Morningside College in Sioux City, Iowa. A new faculty position took Bill back to Southern Utah University where he now works as Associate Professor of Biology, teaching courses in general biology, zoology, and conservation biology. With research interests spanning biological pedagogy to reptile and amphibian community ecology, Dr. Heyborne regularly works with undergraduates on collaborative projects. He and his students are currently studying invasive exotic turtles in the Virgin River, the effect of pinyon/juniper management on reptile and amphibian communities, and the population ecology of an imperiled tiger beetle endemic to southern Utah.

Dr. Lisa Nagaoka is a Professor in the Department of Geography and the Environment at the University of North Texas where she teaches Applied (Urban) Biogeography, Sustainability, a sophomore level course in research methods and career preparation, and coordinates the Earth Science labs. She has broad research interests in human-environment interactions across time and space. Her archaeological research includes estimating prehistoric soil moisture variability across the arid landscape of the Mesa Verde region and its impact on agricultural productivity, documenting changes in the relationship between human foraging practices and resource availability, and evaluating the role of people in megafaunal extinctions. She also conducts sustainability research with undergraduates on campus that includes evaluating the effectiveness of the recycling program, improving bicycle infrastructure to encourage bike community, and documenting native tree distribution to encourage better landscape management practices.

Dr. Elizabeth Olson is an Associate Professor of Anthropology at Southern Utah University with research interests in ethnomedical systems, community-based development, global health, and global environmental change. She enjoys teaching classes that bring students into engaged and hands-on learning experiences—whether locally or abroad. Her classes center on themes such as public health, society and science, field research methods, sustainable development, ecology and health, and anything related to contemporary human lived-experience. Her first book, *Indigenous Knowledge and Development: Livelihoods, health experiences and medicinal plant knowledge in a Mexican Biosphere Reserve* (2014), addresses topics of governance and conservation, ecotourism, and traditional medicine in the context of community development and conservation in the Biosphere Reserve. She has also co-edited with J. Richard Stepp *Plants and Health: New Perspectives on the Health-Environment-Plant Nexus* (2016). Dr. Olson participates in numerous social justice and community-based conservation initiatives in the USA, Mexico, and Guatemala.

Dr. Marsha Bogar Quinlan is an Associate Professor of Anthropology, Affiliate of the Allen School for Global Animal Health, and Honors College Distinguished Professor at Washington State University. She is a medical anthropologist and ethnobiologist with primary concentrations in family health, ethnomedicine, and local ecological knowledge of plants and animals with particular focus on health. She has conducted research with several societies of East Africa and the Americas and was a founding Co-Editor of *Contributions in Ethnobiology* Monograph Series.

Dr. Daniela J. Shebitz is the Executive Director and Associate Professor of the School of Environmental and Sustainability Sciences, Kean University in Union, NJ where she teaches courses in Conservation Biology, Medicinal Botany, Field Biology, Introduction to Environmental Science, and the Senior Capstone that is presented in this edition. She is a plant ecologist whose research interests involve the restoration of culturally-significant plants to traditionally managed ecosystems including beargrass savannas of the Pacific Northwest, the Atlantic white-cedar swamps of the New Jersey Pine Barrens, and medicinal plants in the lowland wet secondary forests of Costa Rica. She has been a part of the Society of Ethnobiology since she was an undergraduate student in 1997 and is incredibly grateful to the Society for helping to shape her professionally and personally. Daniela received the Society of Ethnobiology's Mentor Award in 2017.

Nametha Srikanthan is a graduate with a bachelor's degree in Psychology from York University in Toronto, Ontario. She is currently a nursing student at the University of Saskatchewan. She is interested in research that contributes to changes and advancement in the science curriculum at the university level. Another area of interest is visual arts and art therapy. In the future, she would like to work with autistic children and combine art to promote mental growth. Being an artist, she has participated in art shows and has showcased her paintings as well.

John Richard Stepp is UF Research Foundation Professor of Anthropology at the University of Florida. He is a core faculty member of the Tropical Conservation and Development Program and the Land Use and Environmental Change Institute. He has conducted conservation research over the last two decades throughout the tropics, especially in the Maya Forest and in the Greater Mekong Region of Southeast Asia. His research explores persistence, change and variation of traditional ecological knowledge and ethnobiology. Much of this work has focused on wild food plants and medicinal plants. His work has also focused on patterns and causes in the distribution of biological and cultural diversity. He is a previous editor-in-chief of the *Journal of Ethnobiology*, past-president of the Society for Economic Botany and current president of the International Society of Ethnobiology.

Amanda M. Thiel is currently working towards a PhD in anthropology at Washington State University. Her interests focus on cultural models, medical ethnobiology and Latin America, with present work in Maya communities. She has conducted research in Mexico, Guatemala, and the US. She is also an herbalist and owner of Isla Botanica, a Pacific Northwest-based herbal products company.

Dr. Steve Wolverton is Professor in the Department of Geography and the Environment at the University of North Texas, specializing in archaeology, historical ecology, and environmental geography. Steve is the author of *Becoming a Researcher: Making the Transition to Graduate School* and a co-author of *Applied Zooarchaeology: Five Case Studies* with Lisa Nagaoka (University of North Texas) and Torben Rick (Smithsonian Institution). He is co-editor of *Conservation Biology and Applied Zooarchaeology* with R. Lee Lyman (University of Missouri) and *Sushi in Cortez: Interdisciplinary Essays on Mesa Verde* with J. David Taylor (Stony Brook University). He served as co-editor of the *Journal of Ethnobiology* from 2016 to 2019 and was a founding editor of *Ethnobiology Letters*. In 2016, Steve received the Society of Ethnobiology's Mentor Award.

Dr. Vincent Ziffle is an Assistant Professor of Chemistry at First Nations University of Canada (FNUniv) in Regina, Saskatchewan. He is part of the Indigenous Science program and has interest in the uses and molecular makeup of medicinal plants of the Southern Saskatchewan prairie. This work is made possible by the assistance of First Nations Elders and other Traditional Knowledge holders that guide medicinal plant teachings at FNUniv and surrounding communities. He is a synthetic organic chemist with an interest in food science, and has created a Chemistry of Food and Cooking course with innovative food lab and reliance on Indigenous Food Traditions.

Made in United States
Orlando, FL
05 November 2026

90200709R10114